SpringerBriefs in Operations Management

Series Editor

Suresh P. Sethi, The University of Texas at Dallas
Texas, USA

SpringerBriefs present concise summaries of cutting-edge research and practical applications across a wide spectrum of fields. Featuring compact volumes of 50 to 125 pages, the series covers a range of content from professional to academic. Typical topics might include:

- A timely report of state-of-the art analytical techniques
- A bridge between new research results, as published in journal articles, and a contextual literature review
- A snapshot of a hot or emerging topic
- An in-depth case study or clinical example
- A presentation of core concepts that students must understand in order to make independent contributions

SpringerBriefs in Operations Management showcase emerging theory, empirical research, and practical application in the various areas of operations management (OM), supply chain management (SCM), germane elements of Operations Research (optimization, stochastic modeling, inventory control, etc.) and all related areas of Decision Science and Analytics as applied to the practice of OM, from a global author community. Briefs are characterized by fast, global electronic dissemination, standard publishing contracts, standardized manuscript preparation and formatting guidelines, and expedited production schedules.

Tiziano Pavanini

Rural Demand Responsive Transport

Current Developments and Analysis of a Case Study in an Italian Inner Area

 Springer

Tiziano Pavanini
Politecnico di Milano
Milan, Italy

ISSN 2365-8320 ISSN 2365-8339 (electronic)
SpringerBriefs in Operations Management
ISBN 978-3-031-91394-5 ISBN 978-3-031-91395-2 (eBook)
https://doi.org/10.1007/978-3-031-91395-2

This Springer imprint is published by the registered company Springer Nature Switzerland AG
The registered company address is: Gewerbestrasse 11, 6330 Cham, Switzerland

If disposing of this product, please recycle the paper.

Competing Interests The author has no competing interests to declare that are relevant to the content of this manuscript.

About the Book

Over time, a car-centered mobility system has contributed to the negative externalities that can be observed today in both urban and rural areas: congestion of transport infrastructures, air and noise pollution, reduced urban space for pedestrians and cyclists, lack of parks, etc. All this has led policy makers to find solutions to shift citizens from cars to public transport and other sustainable modes (walking and cycling). While in urban areas, traditional public transport is often an already widespread and effective service that only needs to be transformed to build user confidence, in rural and mountainous contexts, where low transport demand and long distances make it economically unviable, it needs to be integrated or completely replaced by innovative forms of mobility. One of the most valid solutions in this regard is on-demand transport technology, which allows transport providers to reduce their costs by rationalizing the supply (e.g. higher vehicles' load factor) and population of these areas to improve their accessibility to public transport and abandon the use of car. Over time, research has been done extensively in academic literature on the application of this technology in urban areas, but little has been undertaken in rural contexts: this work aims to contribute to research in this field by studying the technical characteristics of a specific case study, in order to provide decision-makers with useful information to counter the phenomena of depopulation and economic and social isolation of these territories.

First, this book provides a comprehensive literature review aimed at understanding the strengths and weaknesses of the DRT service in general and in its application in rural areas: from the description of its historical development, the close relationship between the diffusion of this tool and the phases of technological progress emerges. The central chapters of this work deal with the planning of some DRT services in the inner area of Antola-Tigullio (Liguria Region): this work, carried out after an analysis of the socio-demographic data and the travel behavior of the population, helped to identify the best routes, time slots, and target user groups to experiment with DRT service. After 5 months of experimentation, it was possible to carry out an ex-post analysis of the initial results thanks to the data provided by the local Public Transport Authority (PTA), commissioner of the study.

The results of this work, obtained from the study of the literature and the analysis of a single case study, are multiple and provide useful indications to policy makers and transport providers for the implementation of DRT services in hard-to-reach areas with low transport demand, capable of truly satisfying the mobility needs of the inhabitants by favoring the use of public transport and slowing down the processes of depopulation and economic marginalization affecting these contexts.

Contents

Chapter 1
Introduction

1.1 Introduction

Even before the COVID-19 pandemic, the transport sector—particularly urban mobility—was facing some complicated challenges: constantly increasing traffic, limited green spaces available to citizens and the massive use of private vehicles over public and sustainable modes of transport. The advent of aforementioned contagion has, in fact, prompted local administrations to rethink their urban public transport system. While the negative externalities of this contingency are well known, during this historical period, many growth opportunities have also arisen that policymakers and institutions should seize.

The main mobility trends in place, developed because of the pandemic period and the related restrictions, are mainly attributable to phenomena such as the affirmation of e-commerce and telework and to the repopulation of rural areas in the surroundings of cities.

The repopulation of the so-called low-demand areas—contexts (urban or interurban) with "low or medium-low demand for transport and characterized by a considerable spatial and temporal dispersion" (Campisi et al., 2021), as a result of the need of people to pursue a greater quality of life, occurs after decades of progressive depopulation of the same areas, mainly due to scarce and infrequent transport links. Here, service lines are often neglected by public transport authorities (PTAs) because of the lack of economic sustainability resulting from low transport demand in these territories. Due to an unsatisfactory transport service, many people have been forced in recent years to commute with their own means of transport or move directly to live in city centres in order to avoid social exclusion, marginalization from essential activities and economic impoverishment (König & Grippenkoven, 2020).

T. Pavanini, *Rural Demand Responsive Transport*, SpringerBriefs in Operations Management, https://doi.org/10.1007/978-3-031-91395-2_1

1.1.1 Overview of the Work

Although Demand Responsive Transport (DRT) has already existed for several years, a large part of the academic literature has focused on its application in urban contexts. At a rural and mountain level—territories characterized by a scarce population and low transport demand—little attention has been paid. At the Italian level, as well as in the European Union, there is the firm will to counter the phenomenon of depopulation of rural areas and to guarantee the resident population an efficient public transport service as an alternative to the use of cars. To this end, specific political measures have been promoted, aimed at identifying the territories most in need of intervention. This study intends to provide an analysis of a specific inner area (Antola-Tigullio Valleys, in the Liguria region), through the description of its socio-demographic characteristics and the travel behaviour of citizens. On the basis of this information, obtained through a questionnaire submitted to the mayors of the municipalities involved, it was possible to formulate some hypotheses of DRT services, which were subsequently brought to the attention of the PTA commissioning this study. Based on the ex post data provided by the PTA regarding the first DRT pilot launched (Ne, Val Graveglia), an analysis of the service's performance was then carried out, highlighting strengths and weaknesses.

This manuscript consists of seven chapters. This chapter introduces on-demand transport, remarking on the urgency of finding alternative solutions to traditional public transport for social, economic and environmental reasons. Additionally, it provides an overview of the study, the methodology used for the research and the key theory about transport justice.

Chapter 2 briefly describes the alternative forms of mobility currently available worldwide. It then focuses on the technical characteristics of demand responsive transport by providing a series of definitions found in the literature, illustrating the technological functioning and stating the types and business models of DRT service. Furthermore, Chap. 2 describes the historical evolution of on-demand transport starting from the early 1900s to the present day (many European projects are presented). For this purpose, it is interesting to observe the strong link between the diffusion of this service and technological progress over time.

Chapter 3 presents a substantial literature review of DRT, also stating the successes and failures of the service at a global level and the costs associated with this technology. Chapter 4 describes the characteristics of inner areas and how the Italian Government intends to address, as mentioned, the problem of depopulation in these territories and the social isolation of residents through the National Strategy for Inner Areas (SNAI). In addition, this section outlines the socio-demographic and mobility characteristics of the inhabitants of the Antola-Tigullio inner area. This study was preparatory to the planning of three DRT services in this area proposed in Chap. 5. Chapter 5 reports on the analysis of the mayors' responses: based on these, it was possible to propose three different DRT cases in as many rural contexts. This chapter also presents the inspection carried out on the field and a SWOT analysis aimed at highlighting the strengths, weaknesses, opportunities and threats of the hypothesized DRT services.

Chapter 6 presents an ex post analysis of the performance of the first DRT service implemented in the area (Ne, Val Graveglia), based on the data provided by the PTA. From this study, it was possible to evaluate the efficiency of the service and formulate any corrections. Finally, Chap. 7 contains the conclusions of this work, outlining research limitations and future agenda.

1.1.2 Research Methodology

In this section, the research methodology adopted in this study is presented, aimed at addressing the following research questions:

- RQ1: "What are the main strengths and weaknesses of DRT in rural areas?"
- RQ2: "What are the perceptions of Mayors of inner areas concerning the characteristics of the territory and the residents' travel behaviour with regard to the implementation of DRT services?"
- RQ3: "How did the DRT service perform when tested in the Antola-Tigullio Valley?"

To answer these questions, three different methods were used in this study, depending on the type of data required:

1. Research Method (RQ1)—International literature review
2. Research Method (RQ2)—Survey of mayors
3. Research Method (RQ3)—Data analysis of a specific DRT pilot project

To answer RQ1, an international literature review was conducted (Chap. 3), aimed at identifying the success and failure factors of some of the most important DRT case studies at a global level. From this analysis, it was possible to outline clearly the strengths and weaknesses of this technology.

RQ2 is addressed in Chap. 5 of this work. In the design phase of the DRT services for the Antola-Tigullio Valley, it was crucial to thoroughly analyze the travel habits of citizens and the demographic, orographic and technical characteristics of the territory. In this regard, a questionnaire of nine questions was submitted to the mayors of all the municipalities involved in the project, requesting the following information:

- Current situation of the municipality (population, accessibility and essential services) Categories of users most in need of the DRT service and time slots not covered by the existing traditional transport
- Interventions related to the current traditional transport considered a priority
- Origins and destinations where the DRT service would be more useful
- Possible collaborations with private stakeholders

- Need to transport additional items such as food, medicine and mail
- Involvement of the tourism sector in the DRT service project
- Assessment of residents' availability and willingness to use info-telematic technologies
- Further suggestions

Since the goal of the research was to gather the most impartial and accurate information about the area, it was decided to interview only the mayors of the municipalities involved and exclude the general public. As representatives of their communities, the mayors were the best choice of available options.

However, it should be noted that this choice also presents limitations: although the mayors have excellent knowledge of local transit service provision and gaps in their jurisdictions, they are not necessarily transport experts, nor are they likely users of the DRT services.

A questionnaire-based approach was employed to collect responses that were as uniform and comparable as possible.

For the purposes of this research, a geographic information system (GIS) accessibility analysis was not conducted on the territory as the DRT service proposed integrates/replaces existing lines of traditional transport and for which it was deemed to already have a good socio-demographic and mobility database.

The first DRT service launched in the reference area was analyzed in Chap. 6 to address RQ3 and evaluate the efficiency and attractiveness of the service. Based on the operational data collected and generously provided by the transport manager (AMT S.p.A.), the following aspects of service performances were evaluated:

- Booking method
- Ride duration
- Stop usage frequency
- Booking requests
- Ride requests
- Service time
- Users transported

The analysis of this data helped to clarify the benefits and drawbacks of the service under test.

Figure 1.1 graphically shows the structure of this work indicating the link between each research question identified, the method chosen to address the issue and the chapter of the study in which the results can be found.

Further information on research methods 1–3 is provided in each of the relevant chapters.

1.1.3   Key Theory on Transport Justice

This section of the study sheds light on a key aspect underlying the decision-makers' decisions to implement the DRT service: the concept of transport justice.

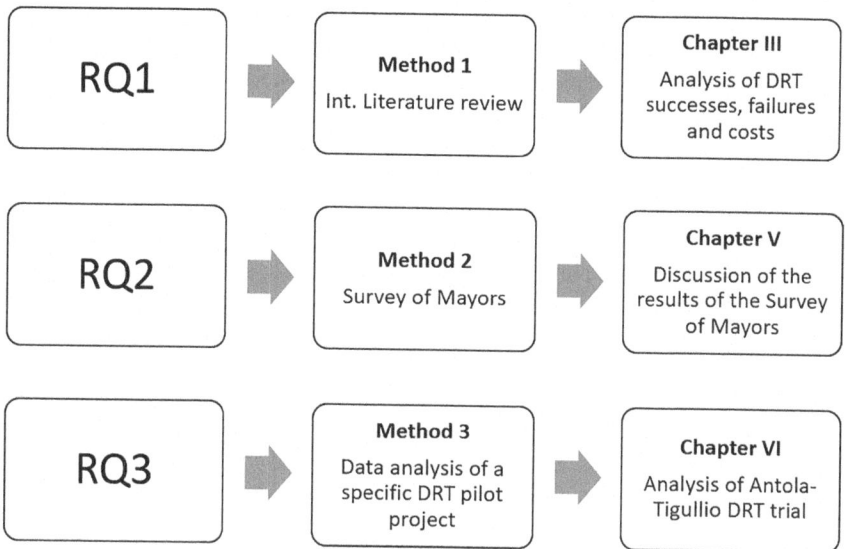

Fig. 1.1 Layout of this study

Social Justice Theories Affecting Transport Scholars

Scholars have adapted main theories of social justice to the transportation industry. The most influential among these are *A Theory of Justice* (1971) by American philosopher John Rawls, *Spheres of Justice* (1983) by political philosopher Michael Walzer and the *Capability Approach* (1980s), a product of the collaboration between economist Amartya Sen and philosopher Martha Nussbaum.

A fair society, in the opinion of Rawls (1971), is one in which everyone has access to the same fundamental liberties and opportunities, and in which social and economic inequalities are structured to benefit the disadvantaged members of society. In Rawls' theory, which has been influential in political philosophy, a wide variety of social issues are addressed such as healthcare, education and international justice. The theory is criticized for being overly utopian and for failing to fully address the problem of dominance and power in society. However, Rawls' theory continues to make a significant contribution to the discussion of social justice and the function of the state in fostering a fair society.

As stated by Walzer (1983), distributive justice should be viewed as an elaborate concept, with various resources and commodities being distributed in accordance with different principles of justice. He distinguishes many "spheres" of social existence, each with its own set of resources and benefits.

The first category of justice, referring to goods and resources traded on the market (e.g. money, property and labour), is the sphere of market goods. According to the author, the distribution rule in this area should be based on merit and effort, such that those who put in more effort and make more contributions to society are entitled to a larger portion of these benefits.

Secondly, the sphere of social goods covers goods and resources distributed through social institutions (e.g. education, healthcare and welfare): the distribution principle should be based on need so that individuals who need these resources and things the most receive the largest part.

Thirdly, the sphere of political power includes the distribution of political influence and power in society. Participation and representation should serve as the foundation in this context so that every member of society has an equal voice in choices that have an impact on them.

Lastly, the sphere of culture covers the distribution of cultural products and resources like music, books and artwork. The distributional ratio ought to be founded on pluralism and variety, allowing various ethnic groups to preserve their own traditions and beliefs.

In political philosophy, Walzer's theory of spheres of justice is important and has been used to address a wide range of social challenges, including transport, education and sustainability. His approach has been criticized for lacking a clear framework for resolving disputes between the many spheres of justice, and for failing to adequately address the problem of allocating resources between them. Nevertheless, Walzer's theory continues to make an important contribution to the current debate on distributive justice.

The "capability approach" is a conceptual framework developed by philosopher Martha Nussbaum and economist Amartya Sen to promote a deeper understanding of social justice and human flourishing. It focuses on people's capacities rather than their access to resources, and emphasizes the value of their ability to achieve their goals and lead satisfying lives. Put simply, the capability approach emphasizes the opportunities and possibilities that people have to pursue their own goals and lead fulfilling lives.

According to this method, people's well-being should be measured in terms of their ability to engage in certain "capabilities" or "functionings" that are necessary for a fulfilling life. Capabilities refer to the actual opportunities that people need to take advantage of in order to achieve worthwhile outcomes, such as being healthy, having access to education and participating in political life. On the other hand, "functionings" are the actual outcomes or states of being, such as being educated or healthy, which result from using these capabilities.

The capability approach emphasizes the importance of considering a wide range of elements, including social, economic and political circumstances that contribute to people's capabilities. It also emphasizes the value of recognizing the diversity of human beliefs and aspirations, and the need to foster a diverse society in which everyone is free to pursue their own goals.

Several social concerns, including poverty, gender inequality and the rights of people with disabilities, have been addressed through the capability approach. It has been used to argue against standard economic development indicators that only include GDP and income, and to support policies that provide social and economic opportunities for neglected people.

The emphasis of the approach on people's autonomy and its understanding of the importance of social, economic and political factors in promoting well-being are some of its key strengths. According to its critics, the approach can be difficult to put into practice and may not provide a clear direction for policy formulation. Nevertheless, the theory continues to make an important contribution to the contemporary debate on social justice and human flourishing.

Transport Justice

Transport justice refers to the fair and equitable distribution of the benefits and disadvantages of transport across populations, regions and times. Infrastructure and transport networks can either improve or limit people's access to essential services, employment opportunities and social activities. As we have seen, this has significant implications for social justice.

Transport justice can be understood along several dimensions: First of all, distributional justice, which involves the fair distribution of resources, benefits and liabilities among various people and regions, including access to transportation services, costs and the geographic distribution of transportation infrastructure.

Secondly, procedural justice, which investigates the fairness and openness of decision-making procedures pertaining to transportation planning and policy, covering topics such as accountability, representation and public engagement. Thirdly, recognition justice, which is concerned with the respect and acknowledgement of many cultural identities, ways of living and preferences, might affect how accessible and popular transportation alternatives are. Furthermore, environmental justice refers to the equitable distribution of environmental advantages and liabilities related to transportation, including problems with air and noise pollution, climate change and resource consumption.

As a complex and multidimensional concept, transport equity often involves trade-offs and conflicts between its various aspects. For example, promoting accessibility and distributive justice through transport infrastructure development could have negative environmental or social impacts. Achieving transport justice therefore requires a comprehensive and integrated strategy that considers the links between many aspects of justice and the complex trade-offs that exist between them.

Achieving transport justice is difficult for a number of reasons. For instance, the distribution of transport services and infrastructure is often unfairly biased in favour of some communities or groups, leaving others with little or no access to vital opportunities and services, thus perpetuating current social and economic inequalities. Many transport policies and plans are developed without sufficient feedback from citizens, especially disadvantaged and marginalized groups. As a result, policies may be implemented that are insufficiently responsive to the wishes and needs of these groups.

Transport planning involves a number of parties with different and often conflicting interests and values. Promoting social equity may conflict with promoting economic growth or increasing accessibility may conflict with promoting environmental sustainability. For many communities and governments, especially those with limited resources, transport projects often require significant financial and human resources. In addition, it can be difficult to secure the institutional support and political will necessary to achieve equity in transport: political factors or bureaucratic constraints can influence transport policies and plans, thus hampering attempts to promote equity and fairness. Finally, particularly for poor and marginalized populations, detailed and disaggregated data on transport patterns and impacts are often lacking, making it difficult to identify and address transport equity problems.

Addressing these issues requires a collaborative and multidisciplinary strategy involving a wide range of actors from the public and private sectors, civil society and academia. It also requires a commitment to openness, accountability and public participation in the development of transport policies and plans.

In recent years, many authors have contributed to the research in this field by providing various definitions of the concept of "transport justice" (Table 1.1).

In order to promote fairness, equity and sustainability in transport systems, a growing field of study called "transport justice" has emerged. Scholars have approached the idea of transport justice from a variety of perspectives, including mobility justice, transport justice and equity. They have also explored a range of issues and themes relevant to transport justice, such as politics and power, innovation and technology, climate change and resilience, collaboration and co-creation methods. One of the most important scholars in this field is, without a doubt, Karel Martens, author of *Transport Justice* (2017).

Martens, taking advantage of Michael Walzer's "Spheres of Justice" theory, states that "transport good, defined as accessibility, should be distributed in a so-called separate sphere, i.e. independent from the way in which other key goods, like money or power, are allocated." (Martens, 2012).

In his work, the author provides a comprehensive examination of the many facets of justice in the context of transportation. The book examines issues of sustainability, fairness and accessibility while addressing the intricate and multifaceted nature of transportation justice. According to Martens, transport systems have significant effects on social justice since they may make it either easier or harder for individuals to access essential services, employment opportunities and social activities. Further discrepancies may come from different transportation policies and practices' impacts on underprivileged and marginalized populations.

Martens proposes a critical review of the current literature on transport justice to highlight the conceptual difficulties and ethical conundrums that arise when defining and applying justice in this context. He offers a conceptual framework for understanding transport justice that takes into account the diversity of interests and values at play and may guide decisions about policy and planning towards more equitable and long-lasting outcomes.

Martens delves deeper into the myriad facets of transport justice, examining the potential effects that various infrastructure and modes of transport may have on the distribution of opportunities and resources among populations, regions and times. Geographic disparities, social exclusion, environmental degradation and health inequities are just a few of the several sorts of transport injustice the author looks at.

Three crucial dimensions (distribution, recognition and participation) are included in the normative framework for transport justice proposed by Martens. The

Table 1.1 Definitions of "transport justice" identified in the literature

Definition	Source
"Equity (also called justice and fairness) refers to the distribution of impacts (benefits and costs) and whether that distribution is considered fair and appropriate. Transportation planning decisions have large and diverse equity impacts. For example: • *Transport expenditures are a major share of household, business and government spending.* • *The quality of transportation options available affects people's quality of life, and economic and social opportunities.* • *Transport facilities and activities impose various external costs including infrastructure subsidies, congestion delay, crash risk and pollution damages imposed on other people.* • *Transport planning decisions can affect development location and type, and therefore accessibility, land values and local economic activity"*	Litman (2002)
"In current transportation planning practice, distributional goals are either not stated at all, are implied but unclear, or, when stated explicitly, are not based on a well-developed moral argument"	Martens (2012)
"a more fair, equitable distribution of the benefits and disadvantages of transportation interventions"	Jennings (2015)
"Recent efforts to emphasize social equity in transportation are emerging as local, regional, and national governments have required agencies to identify and avoid impacts (disproportionately) to low-income and minority population. The U.S. DOT has identified three strategies to address environmental justice: • *Reduce adverse human health and environmental effects on minority and low-income populations.* • *Include all potentially affected communities in the transportation decision-making process.* • *Ensure that minority and low-income populations receive equitable benefits.* *Some agencies have expanded the concept of EJ (environmental justice) areas to encompass transportation constrained populations, such as households without vehicles, disabled persons, and seniors (age 65+), referred to as transportation justice (TJ) areas. Transportation justice can be referred to as the expansion of environmental justice principles to transportation through investigating mobility, access, and modal opportunity"*	Beiler and Mohammed (2016)
"Transport justice in this paper refers to a political ideal primarily concerned with distributional equality, treating people as equals when resources are transferred or distributed among them. This refers to fairness in the distribution of burdens, risks, access, or valuation of assets between different traffic participants. Transport justice thus refers to an achievement of greater equality or the abolishment of injustices"	Gössling (2016)
"Justice considerations stress disadvantaged populations, with the intent to improve equality with respect to accessibility and mobility"	Hananel and Berechman (2016)
"A transportation system is fair if, and only if, it provides a sufficient level of accessibility to all under most circumstances"	Martens (2016)
"Whether people can get to key services at reasonable cost, in reasonable time and with reasonable ease"	Banister (2018)

(continued)

Table 1.1 (continued)

Definition	Source
"Transportation justice describes a normative condition in which no person or group is disadvantaged by a lack of access to the opportunities they need to lead a meaningful and dignified life. It involves transforming the structures and processes that lead to the inequitable distribution of transportation's multiple externalities (e.g., noise, pollution, visual intrusion, risk of bodily harm, and exposure to law enforcement, among others) across populations and space. Also essential to this notion of transportation justice is that residents and other stakeholders should be able to actively participate in and influence the decisions that affect their lives"	Karner et al. (2020)
"Our working definition of transport justice builds on (…) notion of "transport injustice" as a multi-dimensional construct where space distribution is one of three key dimensions that play a determining role in the fair distribution of accessibility. The cost of urban mobility infrastructures is essential in transport justice because transport infrastructure development is a capital-intensive activity. The development of long-term and efficient financing for transport systems is essential for citizens' well-being"	Guzman et al. (2021)
"Transportation planning and infrastructure decisions often ignore the needs of transportation-disadvantaged populations. This creates inequitable outcomes and results in situations where many cannot meet basic needs for mobility and access. In general, drivers enjoy shorter travel times, greater accessibility, and better employment outcomes than those who use other modes. Cyclists of color are more likely to be killed or injured in crashes and subject to law enforcement than white cyclists, and wealthier populations are more likely to have access to high-quality public transit than low-income residents. Furthermore, transportation's dependence on fossil fuels results in substantial greenhouse gas emissions, driving global climate change. It is well documented that the harms of climate change will fall on society's most vulnerable. Our transportation systems, travel behaviors, and policies are therefore critical sites for advancing and implementing equity and justice ideals—Creating a world where people have true access to the transportation resources, they need to lead meaningful, joyful, fulfilling, and dignified lives"	Karner et al. (2023)
"Transportation justice is an important ethical issue of our time and includes reforming and transforming systems, approaches and processes that lead to inequitable distribution of transportation externalities while providing beneficial access to systems and services through procedural engagement in transportation planning across populations and space"	Panikkar et al. (2023)

Source: Author's elaboration

equal distribution of resources, benefits and burdens among different people and regions is referred to as "distribution". The recognition of different cultural identities, habits and preferences is referred to as "recognition", which can affect how accessible and palatable different transport alternatives are. "Participation" refers to the involvement and empowerment of stakeholders in the decision-making processes that affect their lives, ensuring their opinions and needs are heard and taken into account.

In order to demonstrate how transport equity issues manifest themselves in practice and how they can be addressed through policy and planning interventions, the author also offers case studies and examples from different contexts and regions. The importance of a holistic and integrated approach to transport equity is emphasized, recognizing the links between different dimensions of equity and the complex trade-offs that exist between them.

Additionally, a number of scholars have contributed to the discussion on transport justice.

Karner et al. (2020) state that a switch from transportation equality to justice is required. They draw attention to the necessity of tackling past injustices, present-day power disparities, and laws and institutions that affect transportation. In their study on the connection between transport justice and the allocation of urban space in the Colombian city of Bogotá, Guzman et al. (2021) place a strong emphasis on the value of taking into account the various requirements and interests of multiple stakeholders.

Beiler and Mohammed (2016) develop and apply a transportation justice framework to the case study area of Sullivan County (USA), while Litman (2002), in order to assess transportation equity, states some goals to be included in the transportation planning phase by authorities.

While Gössling (2016) investigates urban transport justice, combining three areas where inequities are clear, namely exposure to dangers and pollution from traffic, allocation of space, and time spent in transit, Jennings (2015), exploring the "recognition" and "distribution" dimensions, reviews the literature and policies pertaining to transport justice in South Africa concluding that different assessments used throughout both the planning and impact evaluation stages (such as quality of life, equity and social cohesion) may result in a more in-depth comprehension of any advantages realized. Banister (2018) analyzes transportation inequity in the UK, whereas Karner et al. (2023) present new viewpoints on transportation justice. The authors were able to summarize the state of the art in 2022 of transport justice in terms of theories, evaluation techniques and research questions thanks to the review of 20 articles. Panikkar et al. (2023) investigate the distribution dimension of transportation justice in Vermont areas with high environmental risk, and the findings of their research show that low-income populations and people of colour face greater difficulty in accessing transportation services, which makes it harder for them to find nutritious meals. A "capability approach" to justice and transportation decision-making is set out by Hananel and Berechman (2016), showing the real-world case study of King County (USA), while paternalism and production difficulties are covered by Vanoutrive and Cooper (2019) in their discussion of transportation justice theory, citing the examples of the Transportation Justice movement (California, USA) and the discussion of "basic accessibility" in Belgium.

Other researchers, studying distribution dimension, have looked at the connections between transport justice and walking, public transportation and accessible networks for those with intellectual disabilities (Sagaris & Tiznado-Aitken, 2018; Adli et al., 2019; van Holstein et al., 2022). Verlinghieri and Schwanen (2020) examine the developing debates surrounding mobility justice and transportation, while in challenging times of global warming and fierce urbanization, mobility justice is a topic Sheller (2018) addresses.

In general, the literature on transport justice emphasizes the importance of promoting fairness, equity and sustainability in transport systems, and the need for multidisciplinary and collaborative approaches to achieve these goals. It emphasizes the need to understand the power dynamics that influence transport policies and practices, and the need for participatory and co-creative approaches that prioritize the diverse demands and interests of all stakeholders.

It should be noted that the most recent studies on DRT are strongly influenced by the work of Martens, with a focus on the three dimensions of the Martens structure.

Demand Responsive Transport is among the best options for improving transportation accessibility for underserved communities and ensuring equal opportunities for all citizens, particularly for those who lack access to a vehicle of their own and run the risk of becoming economically and socially isolated.

This study aims to address the three Martens-identified dimensions of transport justice at the distribution level, allowing fair access to public transportation for residents of rural areas, at the recognition level, considering the differences in mobility needs of locals in various contexts, and at the participatory level, involving all interested parties.

The next chapter focuses specifically on DRT by emphasizing the positive and negative aspects that have been noted in the academic literature and outlining the main causes of both successes and failures.

Chapter 2
Demand Responsive Transport: Characteristics and Historic Evolution

2.1 Demand Responsive Transport: Characteristics and Historic Evolution

The objective of this inquiry, as specified in the introduction, is the study of the characteristics of on-demand transport service in rural areas: before delving into its dynamics, however, it is first necessary to describe all alternative options to fixed transit operating worldwide in order to better understand their potential and operating mechanisms.

2.1.1 Forms of Mobility Alternative to Fixed Transit

For years, new forms of public mobility other than traditional transport have been tested at both urban and rural levels. The main purpose of these experiments is two-fold: policymakers aim at avoiding economic and social isolation of some peripheral and poorly served areas and, at the same time, transit providers want to reach financial sustainability and service efficiency. In addition to this, PTAs, as a result of the development of these types of alternative forms of mobility, can become leaders in the field of innovation and improve their corporate image and reputation in the eyes of customers.

New alternative services must compete with private vehicles and offer users an economic incentive for their use in order to deter ownership and use of cars. Such forms of mobility thus represent a trade-off between the comfort and adaptability features typical of cars and taxis and the cost-effective convenience of traditional public transport (Mageean & Nelson, 2003).

The aforementioned types of mobility, developed over time, are adaptable to different contexts thanks to their flexible characteristics. They are generally used in

T. Pavanini, *Rural Demand Responsive Transport*, SpringerBriefs in Operations Management, https://doi.org/10.1007/978-3-031-91395-2_2

cities (especially in hilly or peripheral areas characterized by a high rate of motor-ization) in order to integrate traditional public transport or in rural areas with a very low population density, especially mountainous territory, and a high ageing rate of the resident population.

Before we turn to contemporary DRT, different technologies of this kind of service have long been in use.

The most relevant are collected in Table 2.1.

Table 2.1 Examples of flexible mobility services (FMSs)

Service	Characteristics	Service model[a]	Users	Subsidy
Dial-a-ride	– On-demand service – Stops on request – Non-predetermined route – Booking required	Many-to-Many (from a plurality of origins to a plurality of destinations)	Low-income, elderly or disabled residents	Often state subsidized
Collective taxi	– On-demand service – Stops on request – Predetermined route – Booking required	Many-to-Many (from a plurality of origins to a plurality of destinations)	Residents, workers	Often state subsidized
Shuttle	– On-demand service – Stops on request – Predetermined route	Few-to-Many (from a few points of origin to a plurality of destinations)	Tourists	Often state subsidized
Jitneys	– Stops on request – Predetermined route – Private operators – Minimal regulation	One-to-Many (from a single point of origin to a plurality of destinations)	Low-income, elderly or disabled residents, workers	Usually no state subsidy
Rental vehicles	– Stops on request – Non-predetermined route – Booking required	One-to-Many (from a single point of origin to a plurality of destinations)	Residents, workers, tourists	Usually no state subsidy

[a]Service models are dealt with in the continuation of this work
Source: Author's elaboration based on Lunardon (2011)

Dial-a-Ride

Dial-a-ride services, the objective of study of this work, are briefly introduced here to be then fully described in the course of the whole book.

A dial-a-ride service, also known as Demand Responsive Transport or Paratransit, is a transport service model, which provides for the use of small vehicles by the transit provider (typically 8- or 14-seater minibuses). This service does not present a predetermined time schedule but the PTA can activate it only in event of a user's request. Furthermore, in the most flexible forms, it perfectly adapts its route and stops to the needs of passengers ("Door-to-door" service).

Collective Taxi

Collective taxi is also a compromise between public and private transport with its own characteristics very similar to those of dial-a-ride (on-demand transport service, bookable stop on request, seat on board bookable in advance is compulsory). The substantial difference is relative to the definition of the route: in dial-a-ride services, it varies from time to time based on users' requests, while in collective taxis it is established in advance during the service-planning phase.

Collective taxi is a transport service carried out by minibuses or taxis, normally managed by private companies: it differs from a common taxi for the possibility of sharing the ride with other passengers at discounted prices. Once the operations centre receives user's reservation with basic data (e.g. origin, destination, time and place of departure), an algorithm matches all passengers' needs in order to make a single trip.

Collective taxis represent one of the most widespread forms of mobility alternative to traditional transport in the world, especially in developing countries where they often replace it entirely (being cheaper and faster). Below are some globally known examples.

Marshrutka

The term "Marshrutka" identifies a traditional collective taxi service typically carried out by 15-seat minibuses and widely spread in Russia, Ukraine and in the former Soviet Republics, especially among students and seniors. Widespread in the Soviet Union since the 1930s, it remains today one of the most popular modes of transport in these countries, representing a valid integration to traditional public transport (Weicker, 2020).

A fundamental feature of this service is the possibility for users to stop the bus at any time along the predetermined route of the vehicle, simply by notifying the driver (Gallo, 2020): in fact, there is no call booking system on board the vehicle, but it is necessary to communicate to the driver "At the next stop, please!"

Daladala, Mabasi, Matatu

There are also several examples of collective taxis currently active on the African continent. In Tanzania, particularly in the most populous city, Dar es Salaam, the so-called service "Daladala", or "Dalla-dalla", is active: it became increasingly popular especially from the 1980s, as a result of the collapse of the local public transport system (Kanyama, 2004). Unlike what happens in Western countries where on-demand service integrates fixed transport (FT) or acts as subsidiary, in many African countries it represents the main form of public transport: the stops are not often predetermined but adapt to the needs of customers in real time (Mfinanga & Madinda, 2016).

In Zanzibar, the collective taxi service is carried out by two different means of transport: the aforementioned Daladala (for shorter journeys in urban areas) and the so-called Mabasi for longer journeys between different cities.

In Kenya, the so-called Matatu carries out the same transport service: 14-seater minibuses travelling on pre-established routes and serving about 87% of passenger market share in Nairobi, the capital (Behrens et al., 2017).

Peseros

Examples of Central American collective taxis are so-called *"Peseros"* (the name derives from the cost of the original ticket of one pesos) of Mexico City, which every day serve millions of passengers: they have developed more and more during the second half of the last century thanks to the flexibility of the service they offer and the possibility for users to stop vehicles at any point on the road, regardless of the location of the fixed stops. The owners of these vehicles, earning based on the number of passengers transported and having mild vehicle capacity limits, tend to get as many users on board as possible, overloading the vehicle and moving much faster than their competitors' traditional buses (Roschlau, 1981).

Songthaews, Dua baris, Lain ka

In Southeast Asia, similar vehicles as above provide on-call transport in the form of collective taxis. In Thailand, for example, the "Songthaews" (literally *two rows*) circulate on the streets of the main cities (pickup covered with two rows of wooden benches at the ends). They are very popular among the less affluent population as they represent the cheapest and, at the same time, most flexible form of mobility (Phun & Yai, 2016). The price of transport and the route are predetermined by the transport company, and users can stop the vehicle with a wave of the hand, get on board, settle on the benches, book the desired stop by pressing a buzzer and pay directly to the driver.

The main difference with traditional transport concerns the absence of fixed stops: passengers can get on and off the vehicle wherever they are.

The same service takes different names depending on the country in which it is provided: *Songthaews* in Thailand and Laos, *Dua baris* in Malaysia, and *Lain Ka* in Myanmar. In Southeast Asia, these paratransit services represent an important component of local public transport, integrating and sometimes replacing the lines of FT.

Other Forms of Mobility

As per Table 1.1, the forms of mobility of shuttle, jitneys and rental vehicles are briefly described below.

Shuttles are an on-call service in which the route is predefined, and users can book stops on board the vehicle. It is a one/few-to-many type of service where the shuttle connects a single origin (e.g. the main city station) or a few origins (e.g. the main city station and central market) with a plurality of destinations (e.g. the addresses of each individual passenger).

The jitney ("jitney" was the original slang name for the $5 ticket price) is a transport service where route is predefined, but time schedule is flexible: jitneys' routes are usually circular and move around large commercial hubs, airports and wide residential areas. This service is aimed mainly at workers and poor people.

Finally, for a specific transport (e.g. a group of workers of a company), people can directly rent means of transport such as a minibus capable of fully meeting travel requests.

The Role of State in Flexible Mobility Services

Depending on the specific setting, the political and institutional frameworks in place, and the type of flexible mobility supply being provided, such as DRT, shared taxis, jitneys, shuttles, or rent vehicles, the role of the state might change. However, the government may assist in the provision of flexible mobility supply in a number of ways:

- FMSs can be regulated by the state to guarantee user safety, credibility and affordability. In doing so, it may be necessary to create fare structures, service zones, and emission and safety regulations for vehicles.
- The expense of providing services in socially essential but financially unproductive places, such as rural or low-density metropolitan areas, might be mitigated by the state by giving subsidies or financial aid to operators of FMSs
- The state can encourage the integration of flexible mobility services with other forms of transportation, such as public transportation, walking, and cycling, by using land-use planning regulations. Designing transportation hubs or interchanges that enable transfers between various types of transportation might be one way to achieve this
- For the purpose of delivering FMSs, the state could establish alliances with businesses in the private sector. For instance, the state may enter into agreements with private operators to offer DRT services in places where there is no access to public transportation or collaborate with community organizations to set up shared-transportation programs for certain populations, such as the elderly or the disabled.

Table 2.2 State challenges in setting FMSs fares

Main goal	Issue	Possible solution
Economic sustainability	FMSs need a substantial investment in equipment, vehicles and technology	It may be necessary for the state to set tariffs that support cost recovery and guarantee the service's long-term viability
Users' affordability	FMSs are frequently intended for underprivileged groups of people who would have trouble accessing cheaper transportation choices	The government may have to set prices that are reasonable for these customers and make sure that everyone may utilize the service
Competition	FMSs compete with already-existing transportation services like taxi or public transportation	In order to ensure the service's long-term viability, the state may need to establish prices that are similar with those of other suppliers
Incentives	To encourage consumers to utilize the service, FMSs may provide incentives like discounts or awards	It may be necessary for the state to set prices which achieve a balance between these incentives and the demands of cost recovery and long-term viability
Fairness	Different user groups, particularly low-income or disadvantaged people, may be affected by FMSs differently	The government may have to set fair prices and make sure that everyone in society may use the service

Source: Author's elaboration

When developing FMSs, fares must be taken into account since they might affect consumers' access to and usability of the service. Setting prices for FMSs might provide the state with a number of difficulties (Table 2.2):

As a whole, implementing FMSs may be difficult for the government, and it necessitates careful planning, organization and regulation to guarantee the service's sustainability, usability and safety.

After a description of the main alternative mobility options utilized across the world (collective taxis, shuttles, jitneys and rental cars), the work's focus shifts to on-demand transportation in the next chapters, as mentioned before.

2.1.2   Demand Responsive Transport

On-demand transport allows local PTAs to make their services more efficient by optimizing vehicles' load-factor: a DRT system well integrated with FT on one side discourages the use of private vehicles in favour of public transport, resulting in pollution and land-use reduction, and at the same time incentives citizens to repopulate some residential areas further away from the city centre but with a higher quality of living.

Table 2.3 shows the main definitions of Demand Responsive Transport over time in the academic literature, in chronological order.

Table 2.3 Definitions of DRT in academic literature

Definition	Source
"TTDC's (Tidewater Transportation District Commission) dial-a-ride paratransit service is a shared-ride taxi service; it is offered to low-density residential areas as a feeder service to motorbus routes and as a substitute for evening and weekend motorbus services. Dial-a-ride paratransit service is generally defined as a shared-vehicle service, which provides door-to door service on demand to a number of travellers with different origins and destinations. Dial-a-ride service is a type of demand-responsive paratransit service, since it is flexible in time and non-scheduled. The request for TTDC's dial-a-ride service is made by telephone"	Talley and Anderson (1986)
"Transportation options that fall between private car and conventional public bus services. It is usually considered to be an option only for less developed countries and for niches like elderly and disabled people"	Bakker and Van der MaaS (1999)
"A wide range of local transport which is complementary to conventional, scheduled passenger transport based on large buses, trams and regional trains. This is usually provided by smaller buses, minibuses, vans, taxis and cars and serves dispersed mobility needs, either during hours of low demand, in areas of low population, or where the target users are dispersed among the general population. These services normally act at a very local level, and are either for the general public, or for specific groups (e.g. disabled and elderly). They provide local mobility, as well as connections to other conventional forms of transportation (e.g. from regular bus network to railway service)"	Ambrosino et al. (2000)
"Demand Responsive Transport (DRT) services provide transport "on demand" from passengers using fleets of vehicles scheduled to pick up and drop off people in accordance with their needs. DRT is an intermediate form of transport, somewhere between bus and taxi which covers a wide range of transport services ranging from less formal community transport through to area-wide service networks"	Mageean and Nelson (2003)
"Public Transport systems between conventional public transport systems with fixed routes, stops and timetables, and personal transport systems are usually called "Paratransit systems" or "Demand Responsive Transport systems"…" Generally, because Paratransit or DRTs provides flexible services responding to each user's demand. *"Public transport systems are operated with fixed stops, routes, and timetables determined by public or private companies, who own vehicles and systems. In addition, these systems are available for public use. Most common modes are bus and railway systems. Personal transport systems are operated with route and schedule determined by each user, whose vehicles are owned by each user. Private auto is common case. DRTs, however, consist of systems and vehicles owned by public or private company, and flexible stops, routes and timetables responding to each user's reservation. Because of it, DRT systems have characteristics of both public transport and personal transport. Therefore, when we classify DRTs among transport systems, more categories are needed"*	Takeuchi et al. (2003)

(continued)

Table 2.3 (continued)

Definition	Source
"The term Demand Responsive Transport has been increasingly applied in the last 10 years to a niche market that replaces and feeds into conventional transport where demand is low and often spread over a large area. A typical working definition of DRT is an intermediate form of public transport, somewhere between a regular service route that uses small low floor buses and variably routed, highly personalised transport services offered by taxis"	Brake et al. (2004)
"Increasingly, conventional bus services do not meet the needs of a large section of the population due to increasing incomes and car ownership levels and the resulting dispersal of activity centres and trip patterns. One solution is public transport systems that can operate effectively with lower and more dispersed patterns of demand than the bus, i.e. paratransit..."	Enoch et al. (2006)
"Demand-response is a transit mode comprised of passenger cars, vans or small busses operating in response to calls from passengers or their agents to the transit operator, who then dispatches a vehicle to pick up the passengers and transport them to their destinations. A demand-response (DR) operation is characterized by the following: (i) the vehicles do not operate over a fixed route or on a fixed schedule except, perhaps, on a temporary basis to satisfy a special need, and (ii) typically, the vehicle may be dispatched to pick up several passengers at different pickup points before taking them to their respective destinations and may even be interrupted in route to these destinations to pick up other passengers. *DRT is also known by other terms, particularly in references and publications from earlier years: dial-a-ride, demand-activated transportation, demand-responsive transportation, dial-a-bus, shared-ride paratransit, flexible-route service, and flexible-transport services"*	KFH (2008)
"Essentially, DRT can be defined as an intermediate and highly flexible mode of transportation giving rise to a wide variety of uses"	Laws et al. (2009)
"Historically, the term Demand Responsive Transport (DRT) services has predominantly related to door-to-door Dial-a-Ride services (sometimes referred to as Special Transport Services e STS or "paratransit") provided by statutory authorities and community groups for restricted usage (usually the disabled and elderly). *Interested users would telephone in their requests some days before they intended to travel and the operator would plan the service manually the day before the trip"*	Nelson et al. (2010)
"Demand-Responsive Transport (DRT) is often referred to as a form of public transport between bus and taxi services involving flexible routing and scheduling of small or medium sized vehicles. This means that the vehicle routes are updated daily or in real time by incorporating information on the demand for transportation. Usually, the customers of a DRT service are required to request and book their trips in advance by placing trip requests including information on the origin and destination of the trip as well as the desired pick-up or drop-off time. The vehicle operator uses this information to provide service that satisfies the passenger needs"	Häme (2013)

(continued)

Table 2.3 (continued)

Definition	Source
"Demand Responsive Transport systems provide flexible transport services in which individual passengers request door-to-door rides by specifying their desired start and end locations. Multiple shuttles (or vans or small buses) service these ride requests in shared-ride mode without fixed routes and schedules. DRT services are more flexible and convenient for passengers than buses since they do not operate on fixed routes and schedules, but are cheaper than taxis due to the higher utilization of transport capacity"	Furuhata et al. (2014)
"Demand Responsive Transport (DRT) or Flexible Transit Services (FTS) are considered to be hybrid transit services that combine characteristics of fixed-route services with those of demand responsive ones. In many cases, the implementation of flexible services in urban and suburban areas has led to several benefits, including increased ridership; more cost effective and integrated service for people with disabilities; flexibility in accommodating demand in combination with fixed route traditional services; ability to serve areas with relatively low density to sustain fixed route services"	Papanikolaou et al. (2017)

Source: Author's elaboration

From the definitions of DRT reported in Table 2.1, it is possible to summarize the following key aspects that characterize this technology:

- Shared vehicle service providing on-demand door-to-door transport
- Service is flexible in time and routes
- Important especially for elderly and disabled user categories
- Usually provided by minibuses or vans
- Service useful during hours/seasons of low transport demand
- Service useful in low transport demand areas
- Intermediate form of public transport between a bus and a taxi
- Flexible service responsive to users' needs

2.1.3 How DRT Technology Works

Traditional on-call transport services are managed directly by an operations centre (also known as "Travel Dispatch Centre", TDC) capable of planning flexible routes based on users' requests: customers communicate their booking directly online or by telephone (e.g. places of departure and arrival, number of seats, etc.) to PTA personnel located at TDC and after that, a terminal on board the bus (so-called "Automated Vehicle Locationing"—AVL or "On-Board Unit"—OBU) allows the exchange of information in real time between the Operations Centre and the driver. In this way, the driver is constantly updated on the itineraries and stops they must make.

The service can be planned in two alternative modalities (Mageean & Nelson, 2003): traditional, defined as "centralized" and managed by physical operators, and computer-based, not requiring human intervention. In the first case, the offer is negotiated between PTA staff and passengers and is particularly suitable in peak times, for the elderly or other specific categories of users. Automated system manages DRT service directly via computer, and is particularly suitable during soft hours and seasonal or night services: this type of planning involves a lower cost due to the absence of physical operators.

2.1.4 Typologies of DRT

Enoch et al. (2004) published a report on the potential of DRT technology as a possible solution to the traditional mobility problems of British transport.

They define four typologies of DRT:

1. *Interchange DRT*: This type of service provides feeder links to fixed transport
2. *Network DRT*: This type of service provides both additional connections to existing service lines or replacement connections for some lines considered not performing in specific areas or time slots
3. *Destination-specific DRT*: This type of service provides connections for specific destinations (e.g. airports)
4. *Substitute DRT*: This type of service completely replaces the traditional fixed transport sometimes only in periods of lower demand.

Table 2.4 shows, reporting excerpts from Enoch et al. (2004), the above-mentioned DRT typologies, providing a general description and concrete examples from the past and present.

2.1.5 DRT Service Models

After carrying out a careful analysis of the transport demand by studying its characteristics, flows and preferences, PTAs should define DRT service models in order to establish how to perform the transport service. This depends on the morphological structure of the territory and on the category of users to be served (students, the elderly, occasional users, commuters, etc.).

Depending on the level of flexibility the PTAs want to offer in light of expected transport demand and the available funds, the service models may vary a lot. The continuation of this paragraph lists DRT service models in increasing levels of flexibility, as observed in the literature (Nelson et al., 2004; Papanikolaou et al., 2017; Mageean & Nelson, 2003).

Fixed Route with Predetermined Stops
Among the service models under examination, the one depicted in Fig. 2.1 is the most rigid and thus the least expensive for the transport company. As with fixed public transport, the route and stops are fixed, but the differences are essentially

Table 2.4 Typologies of DRT

	Description	Examples
Interchange	"The first main category is where DRT is additional capacity in order to provide feeder links to conventional public transport. Typically, this would be a DRT service providing an interchange at a rail station or into a bus route. This is interchange DRT"	– InterConnect, Lincolnshire, UK (still active) – Direct Access Response Transit (DART), Bay Area, San Francisco, California, USA (no more active) – Treintaxi, the Netherlands (active until 2014)
Network	"The second category is where DRT enhances public transport either by providing additional services, or by replacing uneconomic services in a particular place or at certain times. Rather than simply being a feeder into conventional public transport, DRT services can be used to provide additional capacity to conventional public transport by serving new markets or to expanding an existing market"	– Wigglybus, Wiltshire, UK (still active but upgraded) – Public Light Bus, Hong Kong, China (still active) – The Belbus midibus, Flanders, Belgium (still active but extended)
Destination-specific	"Here, DRT modes have been developed to serve particular destinations such as airports or employment locations. Once again, in many cases providing conventional public transport would not be economically feasible. A key element of many of these schemes is the presence of a partnership between a local authority and the 'destination' (e.g. a company, airport operator or whatever)"	– Public Vanpool, King County Metro, Washington State, USA (still active) – Allobus Roissy, Charles de Gaulle Airport, Paris, France (still active) – Supershuttle, Los Angeles International Airport, California, USA (shut down at the end of 2019)
Substitute	"Substitute DRT occurs where a DRT system totally (or substantially) replaces conventional public transport services"	– TAXIBUS, Rimouski, Quebec, Canada (still active) – Community Shuttle, Vancouver, British Columbia, Canada (still active)

Source: Author's elaboration based on Enoch et al. (2004)

two: first, this transport service, being "on demand" and not traditional, takes place only if booked by customers and this allows the transport company not to ride "empty", incurring unnecessary costs. Second, vehicles only stop at stops that users have reserved on board or at bus stops. If no reservations are made, vehicles continue on their route without making unneeded stops.

Fig. 2.1 Predetermined route with bookable fixed stops. Source: Author's elaboration

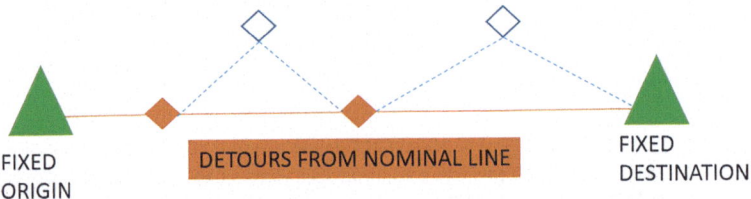

Fig. 2.2 Predetermined route with possible detours from nominal line. Source: Author's elaboration

Fixed Route with Possible Detours

This service model (Fig. 2.2) is more flexible than the previous one: routes are fixed and stops are situated along the main route ("nominal line"). Vehicles can deviate from predetermined routes if users book stops from a location other than fixed itineraries. Particularly in rural and mountainous areas, where the fixed route is frequently identified with the major road and reservations may originate from far-off subsidiary roads, this sort of service model is suitable.

Flexible Route with Fixed Stops

This service model presents a medium degree of flexibility, and the transportation provider plans ahead for some set stops where the vehicle will halt if there are reservations, which are placed at key intersections along the itinerary.

The route, which changes each time according to passengers' requests, is what makes this approach particularly adaptable. Costs for PTAs start to grow here since the service is now more flexible (Fig. 2.3).

Flexible Route with Flexible Stops (Most of Them)

These service models are more flexible and thus more expensive for the transport company.

The "One-to-many" model and its complementary "Many-to-one" model (or "Few-to-many" and "Many-to-few") are two service models frequently used in rural or mountainous settings. Users are transported from a variety of origins to a single destination in the "Many-to-one" scenario (Fig. 2.4) or, in the "Many-to-few" example, to a few selected destinations.

In contrast, the "One-to-many" approach (Fig. 2.5) allows passengers to travel to a variety of locations by picking them up from one or a few sources (in the "Few-to-many" scenario).

Fig. 2.3 Flexible route
with fixed stops. Source:
Author's elaboration

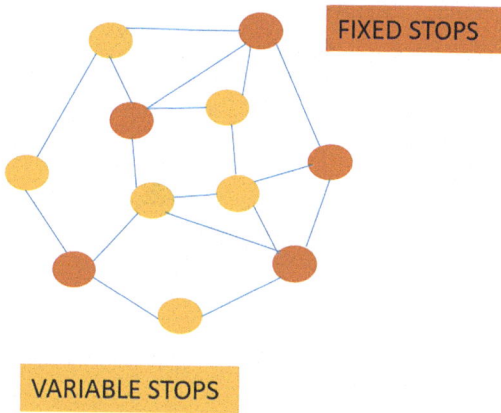

Fig. 2.4 "Many-to-one"
model. Source: Author's
elaboration

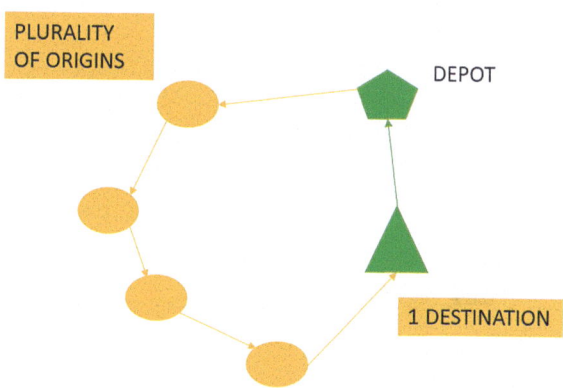

The "Many-to-many" service model, which is typically used for door-to-door transportation of the elderly or disabled (from a variety of origins to a variety of destinations), is the most flexible model found in the literature and, as a result, is the most expensive for PTAs (Fig. 2.6).

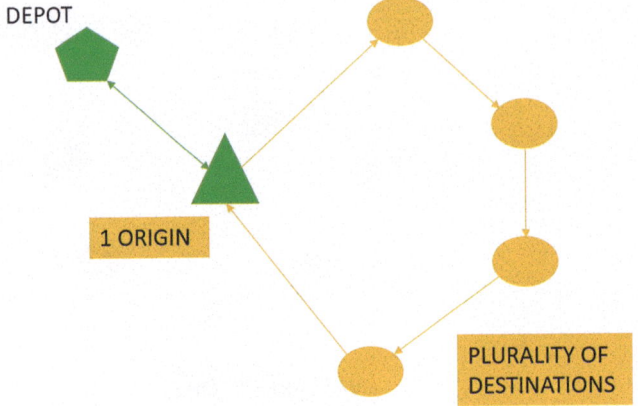

Fig. 2.5 "One-to-many" model. Source: Author's elaboration

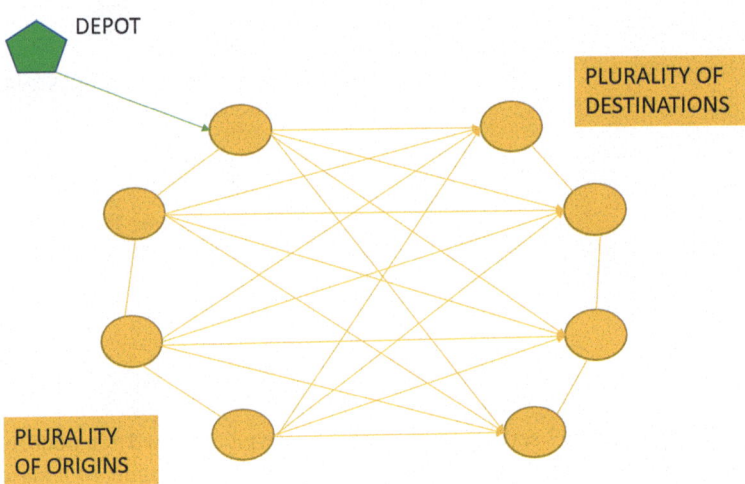

Fig. 2.6 "Many-to-many" model. Source: Author's elaboration

2.1.6 Historic Evolution of DRT

1910–1989

The DRT service, which has become increasingly popular in recent years to remedy some structural problems of urban and rural mobility around the world, presents actually a rather long history that dates back to the early twentieth century (Lave & Mathias, 2009). The first examples of on-call transport spread in the United States with the aforementioned jitneys during the 1910s (the first documented examples date back to 1916 in Atlantic City) (Coutinho et al., 2020): these pioneering DRT services travelled on fixed itineraries and consisted of a series of vans or pickups in which people could get on and off at desired stops.

Due to the willingness of these vehicles' owners to earn as much money as possible, drivers were forced to race at very high speeds to make their service convenient for users compared to conventional transport: due to this, many accidents were recorded at that time on the streets of many American cities (Higgins, 1976).

As reported by the study of Coutinho et al. (2020), the thirty-year period following the appearance of these first examples of prototype DRT (at least until the 1960s) does not record concrete cases of application of this technology, except for some sporadic experimental attempts.

During the 1960s, following the post-Second World War economic boom, in the United States, as in much of the world, many people began to move more and more to large urban centres to seize job opportunities never existed before: this phenomenon, known as *urbanization*, involved large masses of population migrating to cities. People living in the countryside decreased in number and found themselves living in houses very distant from each other and poorly connected at the infrastructural level: these factors represented a problem for the management of public transport in these contexts. It is in such a scenario that the need to think of a different modality of transport, alternative to the traditional system and capable of satisfying the mobility needs of rural citizens, arose: therefore, a cheap on-call transport suitable for particular low-demand transport areas became an alternative or a supplement to fixed public transport (Cole, 1968).

It is in this period that some researchers begin to conduct studies on algorithms capable of planning the routes of vehicles in order to manage transport based on "many-to-many" service models, being able to move people according to a prototype *door-to-door* logic. The most important of these studies is attributable to the CARS (Computer-Aided Routing System) project developed by researchers at the Massachusetts Institute of Technology: Wilson et al. (1969) reported the results of the first study of this kind called *Simulation of a Computer Aided Routing Systems*, in 1969. The authors stated how the CARS project, which involved about 80 researchers, was first sponsored by the US Department of Housing and Urban Environment and later supported by the US Department of Transportation. The main goal of this on-call transport project was immediately clear: to offer a comfortable, flexible and widespread service like taxis at an affordable cost to anyone (like public buses). To do this, the authors understood that the main cost items of traditional transport were represented by number of vehicles and salary of drivers: as a consequence of this analysis, in order to contain costs as much as possible and be able to offer trips at low cost, the natural solution was the reduction both of the number of vehicles employed and of the related number of drivers.

Researchers argued that, through the algorithms developed, they would have been able to allocate each vehicle to travel demands more effectively than a human could ever do. The CARS project was one of the first scientific efforts aimed at studying valid alternatives to traditional transport: although the algorithms developed were not yet fully capable of handling a very large amount of requests, and this required waiting for technological progress for at least another decade, it was still a small step forward in the search for *door-to-door* transport.

The first attempts to remedy the distortions of traditional public transport with forms of on-demand mobility, as mentioned, arose in North America and arrived only in the 1970s in Europe, where some not very successful experiments were recorded.

Starting from the 1980s, on the other hand, with the advancement of technological progress (in particular, the development of the first GPS and GIS systems), the first concrete applications of DRT systems began to be registered in Europe as well.

In Italy, the first DRT service experiment was implemented in the province of Piacenza (Val Nure) at the end of the 1980s (Ferrari, 2018): developed by the Politecnico di Milano, the service was committed to solving the problems of social and economic marginalization of the residents of the mountain valleys. The DRT service made use of 15- to 20-seater minibuses and was configured as a real dial-a-ride service (without a fixed route). Thanks to the success of the trial, the project was confirmed and is still active today, albeit in an updated and expanded form.

1990s

During the 1990s, DRT services spread more and more in Europe, also thanks to important evolutions in the IT field such as new software, better data collection, more advanced algorithms and a wide diffusion of telephones among the population.

Also at the institutional level, this type of sustainable and inclusive mobility solution gained increasing attention, and the European Commission, in 1996, in the context of the Fourth EU Framework Programme, decided to invest in the experimentation of telematics technologies applied to transport: for this reason, the SAMPO (System for Advanced Management of Public Transport Operations) project and its continuation, SAMPLUS (System for Advanced Management of Public Transport Operations Plus), were founded.

The SAMPO Project (Table 2.5)

The SAMPO project, which took place from the beginning of 1996 to the end of 1997, consisted in the evaluation and analysis of some test sites located across the European Union (Finland, Sweden, Italy, Belgium and Ireland) in order to study the

Table 2.5 The SAMPO project

Name of the project	SAMPO
Countries involved	Finland, Sweden, Italy, Belgium and Ireland
Years of experimentation	1996–1997
Means of transport	Buses, coaches, taxis, minibuses, maxi-taxis, etc.
Booking system	Call to the Call Centre
FT Integration/Substitution	Integration
Service model	Different depending on the test site

Source: Author's elaboration

fields of application of telematics processes to Demand Responsive Transport: once categories of users were defined, a step further was to carefully analyze their needs to develop successful on-call services.

The test sites took place in the five EU countries aforementioned. Belgium, Finland and Ireland tested the service in more than one location. One of the main objectives of the SAMPO project was to identify all the stakeholders involved in the development of the on-call service: it should be noted that for all test sites selected by the project, DRT service integrated the existing offer of public transport, thus not providing for forms of substitution but only of addition and integration of FT.

Identifying potential stakeholders is of central importance if aiming at developing a type of service capable of meeting the needs of different actors, each with their own needs and characteristics: in Deliverable D3, an important report written by Finn (1996), an in-depth analysis of users' needs was conducted after their identification and categorization.

This document expressed the importance of a correct identification of the users involved. First, it asserted that a careful study of the potential demand of end users allows identifying the so-called "weak" categories (such as the elderly and disabled) who most need this type of service due to the mobility problems they suffer.

Deliverable D3 stated the need to create, at a European and global level, support technologies for DRT service that are as universal and generic as possible, avoiding individual transport companies having to equip themselves with customized and more expensive technological tools. Furthermore, the list of users identified by this study can be of help to operators and transport authorities around the world: the list provided represented a valid "check list" with categories of users and their needs to refer to.

After identifying the general categories, the SAMPO project researchers decided to investigate the individual categories of users contained herein: for this purpose, the same iterative process described above was used.

This study was conducted by interviewing the main stakeholders involved in the SAMPO test sites, asking them, based on their experience, to rate the degree of relevance of each "User Category" in relation to DRT services, both in general (i.e. at a purely theoretical level) and for the specific test site.

As regards the evaluation of DRT service in general, users were asked to express their opinion by choosing one among the following: "low", "moderate", "high" and "very high".

In relation to the SAMPO sites, each interviewee was asked to assign a value of "1" if the reference "User Category" was considered "significant" for their context, and "2" if it was considered "core". By summing these values across all five test sites, a score ranging from 0 to 10 was obtained, allowing for the comparison of different "User Categories".

From the data obtained, some user categories such as healthcare patients proved to be absolutely well-suited to the use of DRT, while other categories such as foreign operators revealed very little inclination to use this technology.

Based on these data, the SAMPO project researchers selected the user categories most in need of DRT services (those with the best scores) to conduct an even more in-depth analysis of their necessities (part of the successive SAMPLUS project).

Based on the study conducted by Ambrosino and Romanazzo (2002) for ENEA (Body for New Technologies, Energy and the Environment), Table 2.6 shows the main results obtained by the SAMPO project, one of the earliest studies in Europe on user attitudes related to DRT services.

The results of SAMPO project demonstrate that people, regardless of the category of users they belong to, are willing to experiment with new forms of mobility and DRT service, better meeting their mobility needs. Furthermore, the analysis shows how despite on-call service vehicles are smaller than the ones used in traditional transport, thanks to a greater optimization of rides and of space on board, DRT service can be economically advantageous both for transport operators (higher load factor) and for passengers (low ticket fares) (Eloranta, 1998).

The SAMPLUS Project (Table 2.7)

Immediately following the SAMPO project, the so-called "next step" of the process was launched, again thanks to European funds: the SAMPLUS project.

The latter had, like its predecessor, the primary objective of evaluating the on-call transport service and the "readiness to adopt" of some European contexts involved in the project. SAMPLUS included five demonstration sites (of which only one new, the others were already part of SAMPO) and a further four "feasibility studies" conducted in as many locations with structural mobility problems and functional to the study of costs, benefits, problems, advantages and disadvantages of the implementation of a DRT service (Table 2.8).

As stated in the SAMPLUS final report "Systems for the advanced management of public transport", drawn up in March 2000, at the end of the project (1998–1999), by Nick Ayland, the main goals of this initiative were:

- To allow technological transfer between different countries in order to obtain a wide application of DRT
- To conduct feasibility studies in 4 additional test sites in order to allow more countries to implement this service

Table 2.6 Results of SAMPO project

Category of results	Description
Transport policy	At the community level, the main need is to convert private transport to public transport as much as possible, obtaining important results in terms of reducing polluting emissions, reducing traffic and reducing social costs. At national and regional level, the transport authorities must take care of offering an efficient, ecological and inclusive transport service. At the local level, individual municipalities and transport operators must take care of guaranteeing a public transport service to the entire population without marginalizing any category of users.
Users' needs	To be attractive to *passengers*, DRT service should present the following characteristics: • Ease of access • Quick and easy reservations • Last-minute bookings • Reliability of the service • Guaranteed return path • Space on board for luggage • Guaranteed inter-modality with soft mobility • Wide spatial and temporal coverage • Low ticket fares • And more To be attractive to *operators*, the DRT service should present the following: • Maximized load factor • Opening to new markets • Economically convenient service • Technical support systems available • Efficient operations centre • Ability to accept users without reservation • And more
Conflicts	Traditional public transport passengers are used to a service that is not very reliable in terms of punctuality of vehicles and this negative perception of public transport can also be transferred to the new one on call (at least in the early days). Sometimes booking the service can be difficult, especially for categories of users less accustomed to the use of technology. Conflicts may arise between the transport operators and users as the former prefer a booking method managed automatically by the computer while the latter often prefer to speak directly with an operator who incurs a cost for the company. The use of the DRT transport service in rural areas can entice many people to go shopping in larger hubs, rather than in small local shops, which thus see their business decrease.

(continued)

Table 2.6 (continued)

Category of results	Description

Key points	Key points before launching the project:
	  • Careful analysis of the market and of the categories of users to be served
	  • Understand and meet the users' needs
	  • Important service marketing to make the product known
	  • Simple booking procedures
	  • Collaboration between all parties involved
	Key points when the project is active:
	  • Financial coverage for the whole project period
	  • Advanced communication technologies
	  • Constant support service to the user

Source: Author's elaboration based on Ambrosino and Romanazzo (2002)

Table 2.7 The SAMPLUS project

Name of the project	SAMPLUS
Countries involved	Finland, Sweden, Italy, Belgium
Years of experimentation	1998–1999
Means of transport	Buses, coaches, taxis, minibuses, maxi-taxis, etc.
Booking system	Call to the Call Centre
FT Integration/Substitution	Integration
Service model	Different depending on the test site

Source: Author's elaboration

Table 2.8 Demonstration and feasibility sites of SAMPLUS project

	Country/location
Demonstration sites	Belgium/West and East Flanders Finland/Tuusula, Kerava, Järvenpää Italy/Campi Bisenzio, Porta Romana, Florence Sweden/Gothenburg Sweden/Stockholm
Feasibility sites	Finland/Nurmijärvi Ireland/Cavan, Leitrim UK/Surrey UK/West Sussex

Source: Author's elaboration based on Mageean and Nelson (2003)

- To draw up guidelines to allow all stakeholders involved in the project to implement a DRT service in their own context
- To spread the results of this project worldwide to make stakeholders become aware of it

In order to achieve the aforementioned objectives, the SAMPLUS project followed a well-defined methodology, already tested in the previous project SAMPO, which provided for a process divided into five subsequent steps, starting from the analysis of user needs (step 1) and construction of technical specifications of the service in response to these needs (step 2), followed by building a demonstrator and testing it in real cases with passengers (steps 3 and 4) and finally using the results obtained to spread knowledge about DRT to all the stakeholders involved (step 5). The demonstration sites involved went through these five phases, while the feasibility sites only went through steps one and two.

One of the main *lessons learned* from the SAMPLUS project is the necessary cooperation of all stakeholders in order to overcome institutional, legal, organizational and operational obstacles. Often, the main barrier to the implementation of concrete cases of DRT services does not concern technological progress, but obsolete and non-open to change transport policies.

2000s

Between the end of the 1990s and the first decade of the 2000s (EC-funded projects such FAMS), the evolution of DRT changed with the boom of Internet and smartphones, which found increasingly widespread application, not only in Europe but also throughout the world.

The FAMS Project (Table 2.9)

At the beginning of the new millennium, on the basis of the previous experiences of the late 1990s with the SAMPO project and its successor SAMPLUS, the European Union decided to continue investing in the experimentation of this form of collective transport: so, in March 2002, within the EU-IST (Information Society Technology) Program, the European Commission launched the so-called FAMS (Flexible Agency for Collective Demand Responsive Transport Services) project with a dual goal: on the one hand to implement DRT service in two test sites and evaluate their characteristics (as for the SAMPO and SAMPLUS projects) and on the other hand to propose the innovative concept of a "Flexible Agency" to enable a common management for DRT services across Europe. Most of the information available on the FAMS project originated in the FAMS final report entitled "The

Table 2.9 The FAMS project

Name of the project	FAMS
Countries involved	Italy and Scotland
Years of experimentation	2003–2004
Means of transport	Buses, coaches, taxis, minibuses, maxi-taxis, etc.
Booking system	Call to the Call Centre, FAMS web portal
FT Integration/Substitution	Integration
Service model	Different depending on the test site

Source: Author's elaboration

agency for flexible mobility services 'on the move'", drawn up at the end of the project by Ambrosino et al. (2004).

The PTA of Florence (ATAF) was chosen as the coordinator of the entire working group, which included eight partners in total, one of which was a subcontractor.

At the beginning of the 2000s, and largely still today, DRT services applied in real contexts were managed by transport authorities in an uncoordinated way with other stakeholders. Compared to the previous European projects SAMPO and SAMPLUS, the FAMS project set itself the primary objective of taking a step further: that is experimenting in two profoundly different contexts such as Florence (Italy), at an urban level, and the Angus region (Scotland), at a rural level, an innovative form of coordinated management of the on-call service. In this regard, the so-called "Flexible Agency" was conceived, capable of planning and managing the DRT transport of different transport operators, with different fleets, different booking systems and different types of service.

The service provided by this agency can be divided into two categories of recipients: end users in the case of B2C service (Business-to-Consumer) and transport operators in the case of B2B service (Business-to-Business).

The purpose of the agency is therefore to deal with DRT transport managed by many transport operators in different places and times through the provision of an all-inclusive service to make the management of the DRT service as homogeneous as possible despite the presence of many players.

The Flexible Agency conceived by the FAMS project was thus applied in the two selected test sites: in the metropolitan area of Florence, the transport operator ATAF, leader of the whole project, introduced the aforementioned agency in order to coordinate the seven existing DRT services of the area, while in Scotland, in the Angus Region, the Angus Transport Forum tested DRT service for the first time. Table 2.10 reports the main characteristics of the two sites.

The FAMS trial was launched in May 2003 and lasted until February 2004: although the two test sites had different characteristics and levels of preparation, the staff were trained uniformly so trials could be carried out without major problems.

In the final report of the FAMS project, the main outputs obtained were divided into seven categories: "Innovative transport services", "Innovative organizational platform", "Successful take-up and deployment", "Acceptance by personnel and intermediate users", "Acceptance by end-users", "Achieve cost effectiveness and efficiency" and finally "Achieve revenue increases".

As regards the main results achieved in terms of implementation of new services, in general it can be noted that at the end of the FAMS trial in the metropolitan area of Florence, DRT services had achieved a greater degree of cooperation and integration, thanks to the Flexible Agency that was tested. Even in Scotland, where on-call services were implemented from zero, the results obtained allowed the Angus Transport Forum to keep DRT services active even after the FAMS trial (moreover, the degree of involvement of the population was such that around 80% of residents were in favour of maintaining the service).

Table 2.10 FAMS project test sites

Test site	DRT service providers	Total area km²	Total area inhab.
Florence (Metropolitan)	VolainBus CampiDRT Scandicci DRT Sesto DRT Calenzano DRT Porta Romana DRT Disabled	481.89	586.000
Angus reg. (rural)	Glen Esk Services Glen Isla Services Glen Clova Services Angus Disabled and Elderly Services Angus Regular and Special Events Angus Group Hire	2.181	116.040

Source: Author's elaboration based on Ambrosino et al. (2004)

At an organizational level, the results obtained are the most important for the promoters of the FAMS project as the implementation of a Flexible Agency represented the innovative and central aspect of this initiative. In Florence, structural and organizational changes were made to the various existing TDCs in order to make them converge into a single larger one (the FAMS portal introduced, through which passengers could have access to book rides, could now manage up to 300 users per hour, compared to just 20–30 in the past).

In Angus Region, a TDC was introduced where it did not exist before and it received a high degree of appreciation from transport operators both at an operational and organizational level.

As regards the "Successful Take-up and Deployment", reference is made to the degree of reliability and usability of the technologies applied to the new FAMS concept of a single and coordinated TDC: both in Florence and in Scotland, TDC staff greatly appreciated new technologies judging them successful (however, problems could arise in rural contexts with low telephone coverage that can affect the use of the entire system). The use of the FAMS portal for passenger reservations made it possible to reduce a long-lasting problem: "unanswered calls" due to the telephone line blocked by too many calls from users. The FAMS portal has thus made it possible to lighten the number of calls on the telephone line and allow TDC operators to answer more calls than before ("unanswered calls" decreased in Campi Bisenzio).

For example, in Campi Bisenzio, in terms of greater ease of communication between end users and TDC operators, new technologies (such as the FAMS portal) allowed a greater number of connections (360 compared to 30) and a smaller sum of unanswered calls.

As regards the level of acceptance of the FAMS concept introduced in the two test sites, it received high evaluation scores from all the parties involved: operational staff in the TDC, drivers on board the vehicles and transport provider managers. While in Angus Region all stakeholders assigned very high scores to the degree of acceptance of the new technologies, in Florence the scores were high but slightly lower than the pretrial values due to some problems resulting from lack of familiarity with the new tools.

The passengers involved in the surveys on the evaluation of the on-call service were all particularly enthusiastic about the innovations introduced both in Italy and in Scotland: in Florence almost half of the passengers (and a greater share of disabled people) decided to use the service 5 days a week and in particular for home-to-work trips demonstrating loyalty to and confidence in the service.

In Angus Region, although about half of the people own a driving licence, more than half of the residents decided to use the on-call transport proposed in the FAMS project: this is particularly significant information, as in rural areas private vehicle is preferred to the use of public transport, especially in contexts characterized by large gaps in FT network.

The comparison of the costs incurred for the DRT transport can only be studied in the Florence test site, as data were available for the DRT services already existing before the trial: this was not possible in the Angus Region due to the lack of past data.

Therefore, considering the Florentine context, thanks to the technologies applied by the FAMS project, the booking and dispatch cost per passenger decreased by almost 70%. In the same way, both the "operating costs/revenue hour" and the "operating costs/km" reduced.

One of the main criticalities relating to DRT transport service has always been linked to revenues: many cases of application in the past have shown how this type of service guarantees low revenues, insufficient to cover entirely the operative costs and thus DRT has been adopted in regulated market contexts (thanks to state subsidies) rather than in deregulated scenarios where single transport providers were unable economically to meet the costs. The two test sites under examination also showed good participation of stakeholders, a high degree of acceptance by users, but at the same time, rather low revenues at least in the first months of experimentation. The FAMS project has shown how crucial it is to invest in marketing and make the new modes of transport known to the widest possible catchment area: users are often slow to change their transport habits and for this reason, it takes a long time before DRT service becomes the main form of travel for many people. If in Florence some DRT services had already been active for years and this meant that passengers were already accustomed to the use of this technology, in the Angus Region, citizens had to learn about the characteristics of the service from scratch and this, despite the good results achieved, slowed down the spread of DRT.

To conclude, the FAMS project demonstrated how DRT service began to grow in the early 2000s, with more and more users preferring to use public transport over private vehicles.

2010s

Over the last decade, DRT has become a concrete solution for transport providers in order to optimize the load factor of their vehicles and consequently reduce the operating cost of their business. Following the global economic-financial crisis that occurred from 2008, governments have begun to reduce subsidies to traditional transport and, thanks to the simultaneous global spread of smartphones and Internet, the attention of transport authorities towards DRT service increased exponentially: in fact, it allowed PTAs to replace unsustainable traditional transport lines with on-call services designated ad hoc on the needs of passengers.

In the 2010s, there were several projects conducted at international level to evaluate the feasibility and applicability of DRT services: the most important conducted in Europe are presented below.

The LIMIT4WEDA Project (Table 2.11)

As aforementioned, the European Commission intervened again to financially support some sustainable public transport projects aimed at offering an alternative to the use of private cars in some low-demand areas in Europe: for this purpose, within the framework of the MED Interreg program (2007–2013), the EC allocated €1,288,180.00 to the LIMIT4WEDA project (Light Mobility and Information Technologies FOR Weak Demand Areas), for the period 2010–2013.

EC proposed four different categories of actions for these areas (integrated ticketing, infomobility, on-call transport and development of sharing mobility services), and several partners from all over the Mediterranean Region took part in the project (Italy, France, Greece, Malta, Cyprus and Spain) (Campagna & Ambrosi, 2013).

For a better understanding of the text, it is necessary to briefly describe the innovative solutions proposed by the LIMIT4WEDA project (except for DRT service), all suitable for solving the low-demand areas mobility problems.

Integrated ticketing is the possibility granted to passengers to use "one or more modes of transport provided by one or more operators" (Maffii et al., 2012) of a given city/region/country by purchasing a single ticket. In Italy, the cases of BIP (Integrated Piedmont Ticket) should be mentioned in which users, through a contactless smart card, can access all urban and extra-urban public transport and to the regional and metropolitan railway services of Piedmont Region. In Lombardy, the integrated ticket "Io viaggio ovunque", has been active since 2011 and allows

Table 2.11 The LIMIT4WEDA project

Name of the project	LIMIT4WEDA
Countries involved	Italy, France, Greece, Malta, Cyprus and Spain
Years of experimentation	2010–2013
Means of transport	Minibuses
Booking system	Call to the Call Centre
FT Integration/Substitution	Integration
Service model (DRT Perugia)	Many-to-many

Source: Author's elaboration

passengers to use all local public transport services: bus, tram, metro, on-demand services (not managed by private companies), regional trains, boats and vertical systems. Other integrated regional ticketing systems in Italy to mention are UnicoCampania (in Campania Region), Unica Veneto (Veneto Region), Metrebus (Lazio Region) and Mi Muovo (Emilia-Romagna Region). At international level, the most important cases are related to metropolis such as London (Oyster system), Paris (RATP/SNFC system), Melbourne (myki card), Sydney (Opal card) and Brisbane (Go card).

Infomobility allows users to always be updated on discounts, lines and timetables of traditional transport and of new and alternative forms of mobility. Technological development allows infomobility systems to evolve over time and to exploit the platforms most used by users: in North America, for example, Zimride takes advantage of the popularity of Facebook to reach as many people as possible. Zimride is a ridesharing platform that connects students or colleagues allowing them to organize a more efficient transport to the desired destination by sharing the vehicle. Likewise, the Roadsharing platform allows users to share their vehicle in order to optimize travel and parking costs (Dicuonzo, 2020).

Sharing services can assume different nuances (car-sharing, car-pooling, ride-sharing) but all modalities have in common the goal of reducing congestion and air pollution in urban centres through vehicles sharing: passengers can thus enjoy the benefits of a private car (flexibility, capillarity and economic convenience) without the need to own one.

The LIMIT4WEDA project consisted in implementing four trials in selected locations, experimenting the above-mentioned innovations (Fig. 2.7).

Regarding DRT, in Perugia, the local municipality, partner of the LIMIT4WEDA project, decided to experiment an on-call service that could meet the mobility needs

Fig. 2.7 Pilots of LIMIT4WEDA project. Source: Author's elaboration on Google Maps based on Campagna and Ambrosi (2013)

of people in low-demand areas around the city in central Italy. The test site was particularly suitable for a trial of this kind: part of the catchment area was not fully served by FT mainly due to the lack of accessibility (large distance between houses and stops) and the scarce temporal coverage of the service. This led many people to prefer using their own car rather than public transport. For this reason, the municipality of Perugia decided to introduce a completely flexible DRT service (no fixed routes and no fixed timetables) capable of responding to the mobility needs of the residents of these areas. Users could call the toll-free number 800099661 and speak directly with a TDC operator to reserve their seat on board in advance or make a real-time booking (Campagna & Ambrosi, 2013).

The "ProntoBus" service of Perugia attracted more and more passengers, thanks to the greater flexibility of the service in terms of both time and space.

Furthermore, to testify the success of the experiment, from March 2012 (launch of the service) to April 2013, the number of passengers/km for the DRT service doubled (from 1.76 to 2.93), the km/day travelled by minibuses decreased (from 518 km to 350 km), with obvious advantages in terms of cost savings for PTAs. In addition, the cost/km decreased allowing the service to become permanent even after the end of the trial (from €2.03 to €1.59) (Interreg Europe; Regione Lazio, 2013; Campagna & Ambrosi, 2013).

Currently the "ProntoBus" service of the municipality of Perugia can be booked by users through weekly bookings, real time directly with driver or up to 30 minutes before the ride. Nowadays areas served by "ProntoBus" service, almost 10 years after the trial, are the following:

- San Marco
- Cenerente Alta (via Gessara)
- Canneto Alto (Strada della Torraccia, via Macerino, via Piantoni, via Cisternino)
- Canneto Chiesa (Poggio delle Trappole, Str. Canneto)
- Compresso
- Colle Umberto, via Sambuca
- Colle del Cardinale
- Pian di Nese and San Giovanni del Pantano

The Kutsuplus Project (Table 2.12)

At European level, in addition to the projects financed by the European Commission, a series of experiments have been launched in the last decade by city administrations with a view to reducing road congestion and air pollution through the optimization of public transport.

One of the main experiments of DRT service in the European context concerns "Kutsuplus", "apparently the world's first fully automated, real-time demand-responsive public transport service" (Rissanen, 2016).

The Kutsuplus project, jointly conducted by the Regional Transport Authority of Helsinki (HSL) and Split Finland Ltd. and planned for the period 2012–2015, was carried out with fifteen 8-seater minibuses (Jokinen et al., 2019), achieving a great appreciation from the passengers along the whole duration of the trial.

Table 2.12 The
Kutsuplus project

Name of the project	Kutsuplus
Countries involved	Finland
Years of experimentation	2012–2015
Means of transport	Minibuses
Booking system	Call to the Call Centre
FT Integration/Substitution	Integration
Service model	Fixed stops, flexible routes

Source: Author's elaboration

As with all DRT service trials, numerous travel behaviour surveys were conducted before the project to fully understand the travel needs of the local population: in the specific case of Helsinki, the local public transport authority began to conduct this type of analysis since 1966.

Since then, the population of the metropolitan area of the Finnish capital (known as Greater Helsinki) grew from just over 600,000 people to about 1.1 million (in 2012, the year of the start of the trial): due to a growing population, numerous trips by private cars and public transport have also increased, albeit at different rates. For example, trips/weekdays by car grew dramatically from 300,000 in the mid-60s to 1.1 million in 2012 while the use of public transport increased slightly from just under 600,000 trips/weekday to little more than 800,000 (as of 2012) (Rissanen, 2016).

Based on this scenario, it immediately became clear to the Helsinki public transport authority that an intervention aimed at fixing the problems of urban mobility was necessary. For this reason, the authority considered it urgent to implement an on-call public transport service on an experimental basis capable of addressing this issue: HSL occupied the role of project leader and was responsible for the entire trial, planning the service transport and the relationship with all stakeholders. The real on-call transport service was carried out by three different transport operators who took care of making their vehicles, drivers and communication systems available (Rapiditaxi, Taksikuljetus and Andersson). The task of the technology provider, Split Finland Ltd., was to supply and manage the software needed to control the entire transport service.

Based on Kutsuplus final report (Rissanen, 2016), the cost of the ticket assumed for this type of trial was made up of two parts: a fixed base rate and an additional kilometre cost. The average cost for using the service, after an initial test phase with a very attractive price of €1.5 + €0.15/km, following the launch of the service on 3 April 2013, it was increased to €3.5 + €0.45/km: thereafter it has always grown until the end of the project (corresponding to the increase in service quality).

The success of the DRT service implemented in Helsinki by HSL can also be measured by observing the number of trips made by vehicles used in the on-call service from 2012 to 2015: with progressive increase in the number of annual passengers, since the start of the project, the number of annual trips has grown exponentially as well (100,000 trips in 2015 compared to almost 20,000 in the first year of the trial). The small fleet of 15 minibuses almost saturated their capacity because of the increasing demand for transport: this was a problem for HSL, as municipal financial resources to extend the fleet were not enough (Rissanen, 2016).

It is important to observe the costs incurred by the partners of the Kutsuplus trial during the entire project period. Very often, supporting an on-demand transport service involves very high costs for the transport company due to the low-demand areas where transport is needed. The case of Helsinki is no exception and in fact, at the end of the experimentation period in 2015, due to the lack of public finances, the DRT service was suspended.

From the analysis of the financial situation of Kutsuplus project, at the end of each year and in total, what explained above emerges clear: implementing a DRT service can involve a loss of money for the transport authority providing it. This is the reason why often, in the absence of public funding, it becomes unsustainable for PTAs to maintain a service of this type after the end of the test phase.

Observing the data reported in Kutsuplus final report, it should be noted that revenues for each year of service refer to the price of tickets, naturally following the trend of transport demand: improving the quality of service year after year (temporal and spatial coverage) increases the demand for transport and consequently the revenue from tickets. The ticket revenues, in the Kutsuplus trial, have grown exponentially, rising from €2600 in 2012 to €507,700 in 2015 (in 2012–2015, operating revenues amounted to €895,400 in total, including payment of tickets and some minor revenues).

The main cost item of an on-call transport service is represented by the purchase of the transport service from transit providers: this cost follows the annual trend in transport demand as well (the higher the demand, the higher the cost for provision of the DRT service). The so-called operating costs therefore include all the costs necessary to carry out the on-demand transport service: vehicles, drivers, insurance, etc. Other expenses include the purchase and maintenance costs of ICT software and equipment, consultancy from experts and research centres and marketing to make the service known to an ever-increasing number of users. In the case of Helsinki, it is observed that the purchase of the transport service (including "operating costs" and "other expenses") has grown over time as service demand increased: in 2012, the costs for the purchase of service amounted to €316,800 and grew to €3,233,000 in just three years of experimentation (the total cost for the 2012–2015 three-year trial period was €7,821,400).

The salaries of the TDC staff represent other cost items. These costs are more rigid than the aforementioned ones as the number of staff varies little with respect to the increase in transport demand; however, the staff recorded a moderate growth that follows the trend of the service request (personnel costs passed from €119,600 in 2012 to €256,000 in 2015).

To these costs are added a series of minor costs and an annual depreciation (total depreciation €39,100 in the three-year period 2012–2015). From the final analysis of the financial situation of the project, it is possible to note the constant increase of the net cost, which rose from €450,000 in the first year of testing to €2,995,800 in 2015. The net cost of the entire project was €7,913,200 (Weckström et al., 2018): it thus required a substantial intervention from government to cover the costs and return the investment (transport subsidies raised from €200,000 to €2,100,000).

As the demand for transport steadily increased, as mentioned earlier, the capacity of the small fleet of fifteen 8-seater minibuses began to saturate and then HSL, the leader of the project, asked for additional public funds to extend the transport offer to 45 vehicles starting from May 2014 but the relevant authority, in a bad economic situation, refused and, despite the high level of appreciation by stakeholders, Kutsuplus pilot was terminated in 2015.

Summary of Successes and Failures of DRT European Projects

It is useful in this paragraph to summarize the main objectives, barriers, case studies of success and failure of each European project presented above (Table 2.13).

The first European initiative described in this chapter (SAMPO project) had the objective of analyzing citizens' travel habits and studying the most important requirements for the implementation of a DRT service in different geographical areas (Finland, Sweden, Italy, Belgium and Ireland). The researchers thoroughly investigated the main travel needs of citizens as the analysis of their travel preferences represents a key element in the service design phase.

The SAMPLUS project, launched immediately after SAMPO (1998–1999), continued to deepen the study of users' travel habits and the evaluation of DRT pilots in various European test sites (Belgium, Finland, Italy and Sweden). These two projects have contributed together to broaden knowledge on DRT services and on the state of the art of available technologies at the time: results produced by these

Table 2.13 Comparison of European projects

Project	Area	Years	Results
SAMPO	Finland, Sweden, Italy, Belgium and Ireland	1996–1997	Good results useful for extending knowledge on citizens' travel habits and on the state of the art of available technologies: from these outputs, guidelines have been created to implement DRT in other locations as well.
SAMPLUS	Belgium, Finland, Italy, Sweden	1998–1999	
FAMS	Italy and Scotland	2003–2004	The Flexible Agency was accepted by all stakeholders in both test sites. The project has shown some problems of a legal, operational and institutional nature in the adoption of innovative organizational concepts in the public transport sector.
LIMIT4WEDA	Italy, France, Greece, Malta, Cyprus and Spain	2010–2013	The DRT service showed absolutely positive results, deserving confirmation even after the pilot.
KUTSUPLUS	Finland	2012–2015	The pilot had excellent operational results but was suspended at the end of the trial due to lack of funds.

Source: Author's elaboration

initiatives have made it possible to draw up guidelines useful for implementing DRT services in other contexts as well. The FAMS project had the main objective of testing (in Italy and Scotland) a Flexible Agency capable of coordinating all DRT operators involved in a given area, in order to integrate the whole transport network. The initiative involved the city of Florence, which already had several operational DRT services, and the Angus Scottish region, which was instead launching its own project from scratch: in both cases, the test was positive, and the Flexible Agency was accepted by local stakeholders who understood its importance. This initiative has also shown how often the main barriers to the implementation of innovative public transport organization tools are not technological but legislative, institutional and operational. The LIMIT4WEDA project wanted to test innovative mobility solutions in different countries of the Mediterranean area (Italy, France, Greece, Malta, Cyprus and Spain): sharing services, infomobility, car-pooling and DRT. In relation to the latter, a fully flexible ("many-to-many") DRT service was funded in Perugia, Italy. The DRT service showed absolutely positive results, deserving confirmation even after the pilot. Finally, the Kutsuplus project tested the validity of a new DRT service in the city of Helsinki: despite the excellent results obtained during the trial in terms of attracted users, the service was suspended at the end of the pilot due to lack of funds. This initiative shows how often DRT services, without the help of ad hoc public funding, can be economically unsustainable.

The aforementioned European projects have over time raised several issues of the DRT service. The reasons for the success and failure of DRT technology in rural areas, as well as its primary costs, are highlighted in the next chapter's brief international review.

Chapter 3
Demand Responsive Transport: A Short Review

3.1 Demand Responsive Transport: A Short Review

This chapter intends to answer the following research question:

RQ1: "What are the main strengths and weaknesses of DRT in rural areas?"

The paragraphs in this section aim to address this question. First of all, a succinct review of the primary scientific study on the use of DRT technology in low-demand areas was conducted on Google Scholar, Web of Science and Scopus. This was followed by an examination of the major success and failure cases and the identification of major costs in the literature.

3.1.1 Background Research

Demand Responsive Transport offers several advantages for both the society and transport companies, but its utilization remains limited. A study conducted in rural areas of Germany by König and Grippenkoven (2020) examined the underlying reasons and identified psychological factors influencing users' adoption of DRT. The study surveyed 205 families and found that the most significant factors affecting passenger behaviour choices were "Performance Expectancy" (the belief in the system's usefulness), "Facilitating Conditions" (perception of organizational and technical support), and "Attitude towards Car" (propensity for private car use).

The results highlight the importance of "Performance Expectancy" in users' decision to use DRT. Clear communication of the advantages and possibilities offered by DRT is crucial, emphasizing the role of effective marketing strategies. Public Transport Authorities (PTAs) interested in implementing DRT systems

T. Pavanini, *Rural Demand Responsive Transport*, SpringerBriefs in Operations Management, https://doi.org/10.1007/978-3-031-91395-2_3

should promote their services and ensure wide awareness among citizens to encourage higher usage rates.

Another study conducted in Lincolnshire, England, revealed that disabled individuals, commuters and residents of less populated areas are more inclined to switch to using DRT services. Additionally, an ordered logit model applied to a survey in the area indicated that men often used DRT more in retirement than they did while they were working (Wang et al., 2015).

Some studies apply mathematical models to predict demand for DRT services and to optimize routes.

In order to identify the metropolitan areas of the city of Melbourne most prone to the implementation of a DRT service, Jain et al. (2017) decided to estimate the so-called "susceptibility" for DRT in the urban environment: the model used by authors, which is an alternative to the more expensive "user preference survey", studies DRT demand patterns on the basis of demographic characteristics and current population movements. The results of this analysis showed that at an urban level, there are areas that are different from others in terms of their predisposition for the implementation of on-call services; therefore, a model of this type is useful for better understanding in which areas of the city it is necessary to integrate or replace FT with DRT.

Another case of application of a mathematical/statistical model can be found in the industrial area of North Bristol, England. This part of the city, in particular the neighbourhoods of Filton, Avonmouth and Severnside, presented numerous industrial sites and limited parking spaces. It was home to low-income residents and public transport was unable to meet the mobility demand of commuters in the area. Because of this mobility disconnection, the MODLE (Mobility on Demand Laboratory Environment) project was funded to introduce new and innovative mobility systems. The objective of the work of Franco et al. (2020) was, through the use of an agent-based model (ABM) built on MatSim (Multi-Agent Transport Simulation) platform, to build a model capable of predicting the transport demand of two new solutions of experimental mobility: the aforementioned ABM model was able to trace the movements of users through the use of anonymous data extracted from people's mobile phones, provided by Telefonica company. This process was useful in order to trace the best routes of two new on-call services, predict their demand for mobility and measure the advantages that this type of DRT services entail if correctly integrated with traditional public transport and private mobility. The results obtained by the authors demonstrate the validity of the tool used, highlighting its value for policymakers and PTAs who intend to implement an economically sustainable DRT service.

Calabrò et al. (2020) used an agent-based model in order to optimize the routes of a last/first mile service that aims to improve the accessibility of the San Nullo metro station in Catania, Italy. To achieve their goal, the authors used the Ant Colony Optimization (ACO) algorithm, subsequently developed on the NetLogo software. This made it possible to obtain two different routes of 30 minutes of travel time each able to fill the gap in the transportation network.

Papanikolaou et al. (2017) proposed a methodological framework to assist Public Transport Authorities (PTAs) and municipal administrations in evaluating the implementation of DRT services as a supplement to or replacement for FT. The authors sought to discover the critical elements that affect a DRT service's ability to succeed or fail.

The framework consists of four sequential questions that PTAs should ask themselves during the evaluation process. Each question requires different data and inputs for analysis. Table 3.1 provides an illustration of this framework, outlining the specific questions and corresponding information needed. The purpose of this framework is to assist PTAs and municipal administrations in making informed decisions regarding investments in DRT services by considering the relevant factors and data specific to their context.

3.1.2 DRT Successes

Studies comparing DRT with FT consistently show better performance for the first one, as documented by Papanikolaou and Basbas (2021). Coutinho et al. (2020) highlighted the benefits of DRT over conventional public transportation by comparing numerous factors, including the amount of distance travelled, the number of passengers, overall expenses, greenhouse gas emissions and user perception. The authors focused on an urban mobility project in Amsterdam called Mokumflex, which lasted for one year starting in December 2017. The project involved replacing FT in selected areas of the city with an on-demand transport service tailored to the actual needs of residents. Two areas with poor accessibility and limited connections, Amsterdam Zuidoost and Weesp districts, had the DRT service added to an existing FT line (line 49). In Amsterdam Noord, the DRT service completely replaced two previously active traditional transport lines.

Table 3.1 Research questions and related data

	Research question	Input and data
Step 1	In which part of the network should FT be replaced with DRT?	Low frequency/demand/gain service lines or areas with no transport service at all
Step 2	What is the investment required for this intervention?	Choose between full replacement of FT with DRT and integration of existing service
Step 3	In this phase, select the valuation method and define the costs and benefits of the investment.	Choose between cost-benefit analysis and multi-criteria analysis
Step 4	In the last step, define the success or failure criteria of the investment: Compare DRT case with the zero scenario.	Definition of the criteria by which the investment is considered successful or not

Source: Author's elaboration based on Papanikolaou et al. (2017)

The analysis compared the characteristics of traditional fixed transport with DRT. The results indicated a decrease in ridership of DRT vehicles compared to traditional transport. However, the passenger-kilometre driven decreased even more significantly due to the service being specifically designed to meet users' travel needs. Despite the lower ridership, the DRT service enabled PTAs to achieve lower costs due to the reduced passenger-kilometre rate. As a result, the DRT service was confirmed and continued in subsequent years.

The University of Malta, which faces substantial mobility issues because of the high rate of motorization in the country, experienced an examination of the mobility demands of students (Attard et al., 2020). Despite the limited availability of parking spots, a significant number of students (76%) drive their own cars to go to classes. However, studies have indicated that if public transit was of high quality, students who live close to the institution would choose to utilize it. In response, the authors designed and put into action a DRT service with the goals of reducing the number of cars on the road, as well as decreasing traffic congestion and environmental pollution.

Over the course of three days, the authors tested the DRT service nine times with real vehicles while simulating demand for transportation. Using particular key performance indicators (KPIs), they assessed the DRT service's performance. The results showed that government assistance was required since the DRT service cost around twice as much as conventional fixed transit. However, the DRT service showed that it was technologically feasible and garnered favourable student response, indicating that it had a reasonable possibility of succeeding. In the end, the study showed how DRT might be implemented as a way to increase student mobility alternatives at the University of Malta and lessen their reliance on cars.

3.1.3 DRT Failures

Although DRT brings numerous benefits for users and PTAs in terms of greater efficiency, economic savings and flexibility, the list of failures in the world is long.

In the UK, the problems related to traditional transport led PTAs to reflect on possible solutions in order to make transport service more customer-oriented. As a result, DRT technology was chosen as the most appropriate, and Department of Transport and the Greater Manchester Passenger Transport Executive decided in 2002 to commission some British universities to study the best cases of DRT operating globally. This work ended into the so-called "INTERMODE: innovations in Demand Responsive Transport" report produced by Enoch et al. (2006). The authors analyzed in detail 72 DRT cases globally and the most important failures were reported together with *lessons learned* in order to clearly understand the crucial factors determining success or failure of DRT.

The barriers that hinder the full diffusion of the DRT service can be related to three different areas:

- Internal—inside the company (workers, vehicles, service planning, financial management, etc.)
- Micro—everything related to the market in which the company operates (its customers, competitors, suppliers, etc.)
- Macro—relating to what is outside company's own reference market and cannot be directly controlled (global economic and geopolitical situation, social, legal and cultural changes, etc.) (Enoch et al., 2006).

The authors concluded that DRT trials very often fail due to an incorrect or not in-depth a priori investigation of the potential transport demand: sometimes, having not sufficiently studied or predicted the residents transport needs of a selected reference area, PTAs carry the risk of providing an excessively flexible (and therefore quite expensive) transport service or to equip their vehicles with very expensive and unnecessary technological equipment. Furthermore, researchers argued that the marketing aspect of the transport service is also very important for the success of DRT experimentation, in addition to necessary in-house training of new skills, which can be very different from those of traditional transport. Finally, good communication between all the stakeholders involved in the project is essential.

Currie and Fournier (2020) stated that cases of failed DRT implementation experiments in the world are higher than those that become permanent: they argue that the academic literature never focused on DRT failures because supporters' will was to fuel enthusiasm for this innovative form of transport by focusing exclusively on the positive aspects of the service. This in fact allowed the failed DRT projects to be little considered or in any case soon forgotten by the media and public opinion.

To conduct their analysis on DRT failure cases, the researchers looked for information on some specific DRT-dedicated reports, always double-checking results to assess whether services were still active or not; from this long work, they extracted and reported on a single database a total of 120 DRT cases from 19 different countries.

They divided the life cycle of on-call transport into three main phases starting from 1970: the first stage (1970–1984), defined "Early DRT Dial-a-Bus Development", included the first cases of DRT trials characterized by high failure rates. During the second phase of the service history (1985–2009), called "Paratransit/Community Transport DRT Era", in which public subsidies have greatly helped the financial stability of services around the world, the number of trial failures has been quite low and many DRT services have since become permanent. The third phase of the life cycle of DRT, "ICT Tech Micro-Transit DRTs", from 2010 to the present day, despite availability of increasingly advanced technologies, presents a higher failure rate than the previous era (Currie & Fournier, 2020).

From this analysis, a large number of DRT trials still active in 2019 (reference year) were launched between 1997 and 2004 (therefore within the so-called "Paratransit/Community Transport DRT Era"): 58% of DRT trials launched during the second era are still active to this day (updated to 2019) making this period the most successful in DRT history.

From the analysis of the 120 DRT systems collected by the authors, the results showed that in general, regardless of the historical launch period of the trial, DRT services tend more to fail than to become permanent: about half of the cases observed lasted less than 7 years and 40% of them had a maximum lifespan of 3 years.

The authors' aim was to understand reasons behind this phenomenon: the main cause identified is attributable to the relation between high costs of service and high failure rate. The ultimate phase, "ICT Tech Micro-Transit DRTs", revealed to be the least successful due to the use of expensive vehicles and very flexible routes, whereas the "Paratransit/Community Transport DRT Era" presented the lowest costs, as the service was often carried out as a voluntary activity, with drivers working without pay. This naturally implies that complex and very flexible types of services ("many-to-many" models) are also the most expensive for the transport company.

3.1.4 DRT Costs

In Florida, USA, the authorities also decided to analyze DRT service in order to use it as a possible alternative to the traditional transport service. In 2008, the Department of Transportation of the State of Florida commissioned the Centre for Urban Transportation Research of the University of South Florida (Tampa) to produce a report that could identify the main categories of expenses related to DRT service, collecting some best practices to contain costs. In July 2008, Goodwill and Carapella (2008) delivered the final report "Creative Ways to Manage Paratransit Costs" in which the above was treated.

When it comes to on-call service in the United States, a due premise is necessary: the introduction of the *Americans with Disabilities Act* (ADA) of 1990 revolutionized the transport sector in the country. This law prohibits discrimination against people with disabilities and guarantees them equal rights in all contexts of public life such as school, transport, work and areas of public use (libraries, parks, swimming pools, etc.).

In the transport sector, this law forced each fixed-route service provider to introduce in parallel an ADA complementary paratransit service for people with disabilities compliant with ADA requirements. For example, "*ADA complementary paratransit service must be provided during the same service hours as fixed-route service and within a three-fourths mile corridor of the fixed route.*" Furthermore, "*each provider must have an established eligibility process to determine whether a customer who requests use of the ADA complementary paratransit is eligible for such a ride*" (NADTC, 2014).

This law led to profound changes in the corporate structure of American transport companies, forcing them to suddenly enter a business whose dynamics they did not know.

The researchers, in order to identify the main categories of costs associated with the paratransit service provided in Florida, conducted a search in the National

Transit Database (NTD) and in the Florida Commission for the Transportation Disadvantaged Annual Operating Reports (AOR): thanks to this multitude of data, they were able to identify the major cost items and their evolution over time.

The 12 main cost categories identified in 67 different Florida counties (related to the fiscal years from 2002–2003 to 2005–2006) are listed below (each cost category includes many cost items within it):

- Personnel costs
- Fringe benefits
- Services
- Materials and supplies consumed
- Utilities
- Casualty and liability
- Taxes
- Purchased transportation services
- Leases and rentals
- Annual depreciation
- Contributed services
- Other

Annual reports provide information on the composition of the total expenditure of transit providers: the 12 main cost categories have been calculated individually in order to understand how much each of them contributes to the total expenditure. The main cost item is represented by the cost of labour (36%), followed by the purchase cost of the transport service (22%), after which, to a lesser extent, the expenses for materials and supplies (9%), fringe benefits (8%), services (7%), insurance expenses (5%), other expenses (4%), equal annual depreciation and leases/rental (3%) and finally taxes, contributed services and utilities (1%). The evolution of these expenses has been studied during the period 2003–2006: due to stringent financial possibilities by transport providers, each expense item has remained more or less stable over the period of time considered (a slight increase is noted only for the two main cost items, namely "labour" and "purchased transport").

The analysis of NTD Reports is important in order to understand the evolution over time of operating costs and maintenance costs of the service. These reports look at 25 public transport providers in the state of Florida for fiscal years from 1999–2000 to 2004–2005. In general, operating costs are much higher than maintenance costs and represent almost all expenses incurred by transport companies: the "Total Operating Expenses" raised from $66,301 in 2000 to $165,524 in 2005, while the total costs of maintenance grew significantly less (from $11,293 in 2000 to $22,392 in 2005).

The above testifies, in the six-year period considered (1999–2005), a lesser growth for service maintenance costs (98.3%) and very high growth for operating costs (149.7%). The hourly cost of the service also grew exponentially from 2000 to 2005 from $23.31/h to $36.52/h (57% increase).

Table 3.2 shows the main strategies implemented by transport providers in order to keep total costs of DRT service low. These measures were identified by researchers based on the best practices observed.

Table 3.2 Strategies to keep DRT total costs low

Area of interest	Description
Service areas	USA transport providers, since they must comply with ADA requirements, must try to offer a transport service able to meet the needs of any passenger in order to optimize business costs and avoid having two separate services: Fixed route on the one hand and DRT on the other. Another strategy to contain costs concerns the possibility of meeting the ADA requirements at the minimum allowable: For example, limiting the service area of DRT services exclusively to the area within 3/4 mile of the already existing fixed-route service. For DRT services that exceed this space, transport providers must charge higher rates in order to increase revenues
Eligibility process	Another way to keep costs down concerns the possibility of controlling the demand of transport: This can be done through the eligibility process (it defines whether passengers are able to use the fixed-route service or, based on their degree of disability, are entitled to DRT service). In order to better identify people who need DRT service, proposed strategy is to define eligibility application forms that are as precise as possible and that include a medical verification of the passenger's doctor
Recertification	After carrying out the analysis for the eligibility process and accurately identifying their potential demand, transport providers must constantly certify the conditions of their DRT passengers on a regular basis (this allows them to be constantly updated on changes in policies and on the conditions of users about their status, relocations and deaths). The frequency of recertification depends on each individual company, although the report shows three years as the average
Reservation process	After defining the operational area of the service and identifying transport demand, the next step is to obtain an efficient reservation/scheduling/dispatch process. Each transport provider must have an efficient reservation process: The company must communicate as clear as possible the booking window created to customers. It is useful for containing costs to negotiate the exact pickup time directly with passengers so that the company can spread the demand over a wider time slot and users benefit from great flexibility. In order to communicate clearly with passengers, it may be important to provide users with guides on the reservation processes and use of DRT services. Subscriptions to the service allow transport agencies to manage better reservations. This is because subscribed passengers are often commuters who always have the same travel times and the same origin and destination
Scheduling process	The scheduling process is now completely automated but the physical presence of the operators is still very important: a double check by the staff is always useful to avoid system errors. It is important for operators to set up a system that contacts users before the time of their ride. This reduces the so-called "no-shows" and confirms the present passengers
Dispatching process	Some transport providers have systems that provide for direct communication between dispatchers and drivers at each pickup and drop-off location. This allows the dispatch Centre to be constantly updated on changes in the schedule, which, in the event of delays or no-shows of passengers, must be modified (this can only happen in contexts with few operators present; in larger contexts, the driver contacts the dispatch Centre only in case of problems)

(continued)

Table 3.2 (continued)

Area of interest	Description
Use of technology	Investments in technology represent a cost containment solution. The digitization of the booking system, both via app and web, allows the PTAs to save on TDC personnel costs. Furthermore, DRT service is based on the use of so-called automatic vehicle location (AVL) terminals on board vehicles. The greater the technological advancement of these systems, the better the degree of real-time response of the TDC to any problems encountered by the driver. Together with the AVLs, the drivers are also equipped with Mobile data terminals (MDTs), which are terminals with which the driver can record the exact times and locations of passengers' pickup and drop-off
No-shows and late cancellation policies	Some of the main problems related to the planning of DRT services are attributable to the so-called "no-shows" (passengers who simply do not show up at the agreed time and stop without giving any notice) and to "late cancellations" (passengers who decide to contact the customer care of the transport agency to cancel their ride but with insufficient advance to allow TDC to reformulate a new route for the DRT vehicle). Since practices such as no-shows and late cancellations involve heavy losses, the transport operator must impose strict policies on passengers who intend to abuse the system. One solution could be to record the behaviour of individual passengers and provide bonuses to those who behave flawlessly, and penalties to those with high rate of no-shows and late cancellations (e.g. Citifare, in Reno, USA)
Contracting	A possible strategy to contain costs and cope with fluctuations in transport demand lies in the possibility of dividing the management of PTA'S own transport service: Some services are carried out by their own vehicles and with PTA'S own personnel, while contractors are used in other lines
Service monitoring	PTAs must be equipped with service monitoring systems, as this ensures that the quality of the service is always high, prevents any fraud and protects the rights of passengers. These monitoring systems use KPIs to evaluate the performance of the service from all points of view (including user feedback)
Use of fixed routes	As previously mentioned, a key aspect for containing costs is linked to the in-depth study of transport demand. The main objective of transport agencies is to convey as many users as possible to fixed-route services. This can be done in several ways such as better accessibility of information and bus stops
Fixed-route travel training	In order to shift regular users of DRT to fixed-route service, PTAs can use different strategies including offering reduced or free rates as an incentive or providing individual or group training
Use of volunteers	The simplest strategy to contain costs lies in the availability of volunteers able to perform certain roles within the PTA for DRT service, for example, carrying out training sessions for passengers, information services, etc.

Source: Author's elaboration based on Goodwill and Carapella (2008)

3.1.5   Concluding Remarks

In order to answer the first research question of this work, a careful analysis of the international literature has been done in order to pinpoint the key traits of the DRT service in rural regions, the driving forces behind success and failure cases and the key costs related to it.

Although the fundamental elements of this technology are universal, it should be highlighted that any regional, cultural and governmental environment can have a different impact on DRT performances. The Antola-Tigullio inner area case study is covered in the next chapter.

Chapter 4
Italian "National Strategy for Inner Areas" and Analysis of Antola-Tigullio Case Study

4.1 Italian "National Strategy for Inner Areas" and Analysis of Antola-Tigullio Case Study

4.1.1 Inner Areas Description and SNAI Initiative

This chapter discusses the distinctive characteristics of inner areas and the National Strategy for Inner Areas (SNAI), the Italian government's plan to address the depopulation issue of these regions and social isolation of residents. The socio-demographic and mobility characteristics of the people living in the Antola-Tigullio inner area are also presented in this section.

In order to clarify the political measures and concrete actions implemented by public authorities to deal with inner areas, it is useful, first, to refer to the guidelines dictated by Italy's National Strategy for Inner Areas to ensure a brief but representative description of the broader framework within this initiative is inserted.

In so-called "inner areas", there has been population loss in parts of Italy and the EU. These areas are remote from urban hubs that offer necessities such as transportation, healthcare and education. Since the middle of the last century, locals—notably families and students—have been relocating to nearby cities due to the lack of amenities and for employment possibilities. Due to this migration pattern, inner areas are gradually being abandoned, leaving behind an ageing population.

The migration and depopulation of inner areas have social, environmental and economic impacts. Socially, the younger population leaving these regions and the ageing residents place significant strain on the national pension system. In terms of the environment, cultivated areas that are abandoned may experience medium- to long-term hydrogeological instability and an increased danger of landslides. Economically, local businesses close down, and private companies show little interest in investing in these areas, forcing working-age residents to commute long distances for employment.

T. Pavanini, *Rural Demand Responsive Transport*, SpringerBriefs in Operations Management, https://doi.org/10.1007/978-3-031-91395-2_4

To address the depopulation issue, the Italian government introduced the "National Strategy for Inner Areas" in 2014. This plan, which is part of a bigger European initiative, seeks to counter the aforementioned tendencies and create value for these regions. The SNAI strategy has selected only 72 inner areas at the national level in order to concentrate resources and energy on a few case studies without the risk of dispersing them. The objective for each inner area is twofold: first of all, promotion of the territory from an economic and productive point of view and then the creation of essential services to prevent the residents of these territories from migrating towards cities. For the purposes of this initiative, territories with the highest level of criticality but also unexpressed potential were selected.

On 9 December 2013, the draft of Italy's Partnership Agreement on cohesion policy for the period 2014–2020 was sent to the European Commission. This document, which had to be drawn up by each Member State, set out the strategies and procedures with which each State intended to use ESI (European Structural and Investment Funds) resources in order to pursue the common cohesion policy promoted by the European Commission and aimed at sustainable and inclusive community development. The technical paper *National strategy for inner areas: definition, objectives, tools, and governance* has been annexed to the previously mentioned document. It provided a thorough analysis of the status of Italian inner areas by outlining existing standards and development procedures.

Inner areas are often distinguished by a lack of easy access to basic services and by regions rich in unique cultural and natural assets. In addition, because of how significantly different these regions are from one another, it is almost difficult to apply general principles that are not tailored to the particular setting.

Based on above, it is crucial to understand how inner areas have been identified by the government and based on which criteria. As part of this project, it was decided to proceed by radially defining the distance of these areas from the nearest "service provision centre"; each kilometre range is associated with a different degree of accessibility to these poles.

The report *A strategy for inner areas in Italy: definition, objectives, tools and governance* (2014) provides the exact definition of "service provision centre":

> A 'service provision center' is identified as a municipality or group of neighboring municipalities able to provide simultaneously: a full range of secondary education, at least one grade 1 emergency care hospital (DEA)[1] and at least one Silver[2] category railway station. (UVAL, 2014)

---

[1] "Grade 1 emergency care hospitals (DEA) include a set of operational units that, in addition to Casualty departments, guarantee observation facilities, short stays, resuscitation and diagnostic-therapeutic general medical intervention, general surgery, orthopaedics and traumatology, cardiology intensive care. They are also able to provide chemical, clinical and microbiological laboratory services, medical imaging and carry out transfusions." (UVAL, 2014)

[2] Rete Ferroviaria Italiana (RFI), manager of the Italian railway infrastructure, classifies Italian stations into different categories on the basis of size, average number of passengers who frequent the station per day and average number of trains crossing that station per day: the SILVER category identifies small-medium sized stations that guarantee a metropolitan, regional and long-distance transport.

Essential services for residents in inner areas include a comprehensive school system, a hospital with emergency, resuscitation and diagnostic-therapeutic capabilities, as well as a railway station for metropolitan, regional and long-range transportation. The Italian territory is organized in a polycentric manner, with medium to large centres surrounded by smaller settlements that often lack essential services. The "outlying areas" are located approximately 20 minutes away from the attractive pole, and residents can easily access essential services through integrated suburban and urban public transport. The financial, social and transportation conditions for inhabitants diminish as the distance to the main centre increases, as per "intermediate areas" (20–40 minutes), "peripheral areas" and "ultra-peripheral areas" (40–75 minutes or more than 75 minutes). Since PTAs cannot sustainably generate an income from the provision of these services, DRT services may completely replace conventional public transit in specific areas.

The supply of the three services designated as essential (education, health and mobility) is the foundation of this project of economic and social rehabilitation of the inner cities of Italy. To ensure the best use of resources, the current provision of critical services needs to be reconsidered. For these areas, which are characterized by considerable distances between houses and impervious territory, a capillary distribution of such services throughout the area proved to be wholly ineffective.

The above is what happened in recent years for healthcare; in order to rationalize costs and increase the quality of service, the many small clinics distributed throughout the area have been closed and merged into a few larger medical centres. Figure 4.1 illustrates the main measures that SNAI strategy intends to adopt to reorganize healthcare in these territories.

As per education, the aforementioned phenomenon of ageing of the resident population in these areas has led over time to a declining number of students: often the only schools present in the area create unique classes in which students of

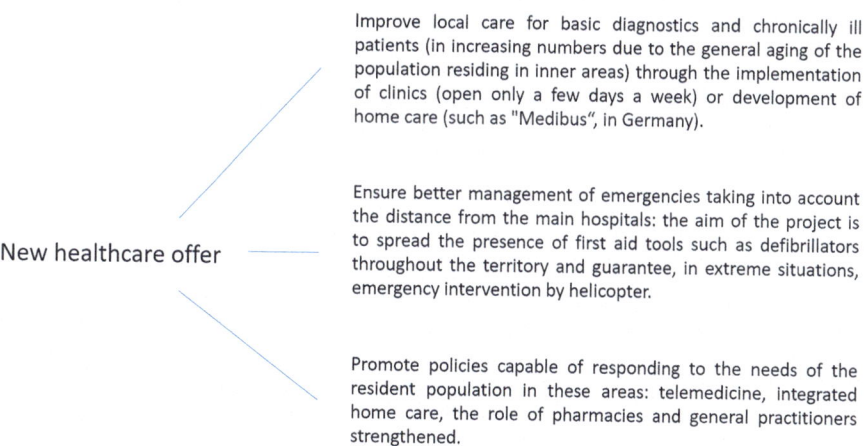

Improve local care for basic diagnostics and chronically ill patients (in increasing numbers due to the general aging of the population residing in inner areas) through the implementation of clinics (open only a few days a week) or development of home care (such as "Medibus", in Germany).

Ensure better management of emergencies taking into account the distance from the main hospitals: the aim of the project is to spread the presence of first aid tools such as defibrillators throughout the territory and guarantee, in extreme situations, emergency intervention by helicopter.

New healthcare offer

Promote policies capable of responding to the needs of the resident population in these areas: telemedicine, integrated home care, the role of pharmacies and general practitioners strengthened.

Fig. 4.1 New healthcare offer. Source: Author's elaboration based on UVAL (2014)

different ages are grouped together, resulting in a rather low quality of education service offered and not suited to the needs of individual schoolchildren.

This situation pushes many families to leave inner areas and move to bigger centres that offer students more opportunities. The SNAI project intends to intervene stopping this process of depopulation: the aim of this project, related to education, is to improve the school services offered to students residing in these areas, avoiding their need to move and strengthening the link with their territory. In terms of education, the project includes the following measures (Fig. 4.2):

After the description of the main measures in terms of education and health, SNAI initiative focuses on the improvement and reorganization of the mobility offer (Fig. 4.3).

The SNAI project also defines the governance model for the measures needed to relaunch Italian inner areas: although this strategy is of a national nature and

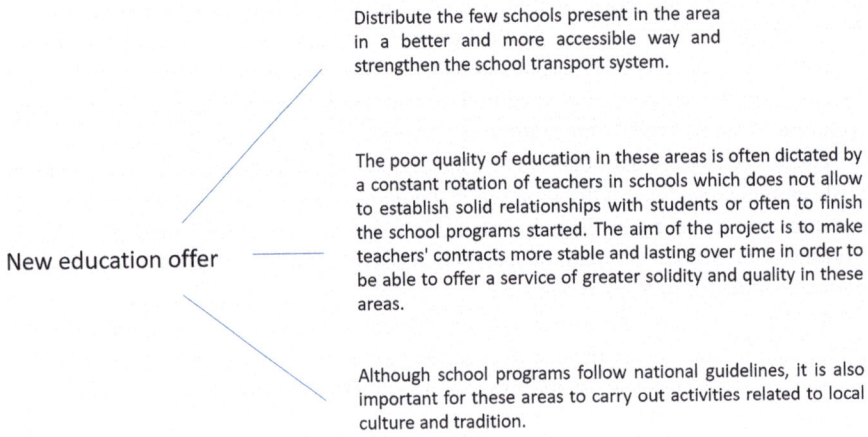

Fig. 4.2 New education offer. Source: Author's elaboration based on UVAL (2014)

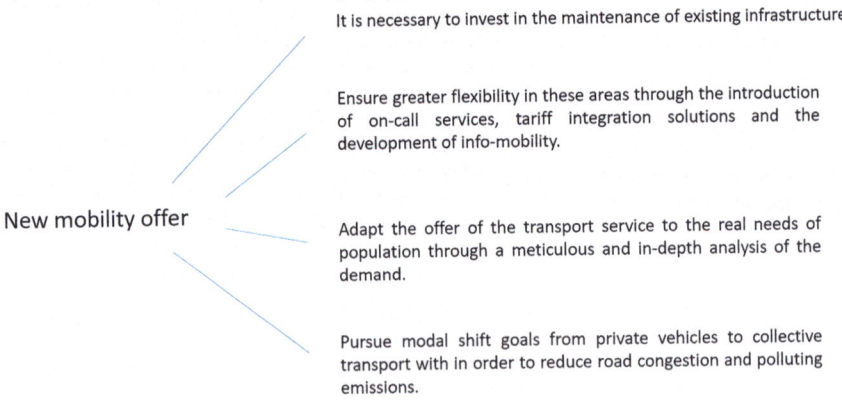

Fig. 4.3 New mobility offer. Source: Author's elaboration based on UVAL (2014)

involves the whole country, the responsibility for identifying these areas and the implementation of the established measures falls on the regions (in the initial phase, a single inner pilot area is chosen by them).

Central Government, on the other hand, had the task of monitoring actions undertaken by Regions during the process: monitoring activity took place with selected indicators that allowed evaluating the impact of measures taken and the success or failure of these initiatives. Government also created a platform for sharing *lessons learned* during the course of this project. This was the purpose of the so-called "Federation of Project Areas".

The SNAI strategy also provided that each measure converge within a so-called "Area Project": each initiative was implemented through a "Project Framework Agreement" (PFA), undersigned by the Government, Regions and Local Bodies and containing the interventions to be realized and the related funds. This agreement enabled communication to all stakeholders involved.

4.1.2   Antola-Tigullio Inner Area: Analysis of the Context

Socio-Demographic and Mobility Characteristics of Antola-Tigullio Valleys

This chapter deals with the central theme of this work: the study and planning of a DRT service in the Antola-Tigullio Valley, a site optioned by Liguria Region as a prototype inner area on which to concentrate resources in the initial phase of the SNAI project.

Antola-Tigullio area, one of the four Ligurian inner areas included in SNAI, is in the province of Genoa and consists of 16 mountain municipalities (area of about 600 sq. km), described as per Table 4.1.

The morphology of this area is particularly complex, as evident from the difference in terms of altitude between some municipalities: Ne, in Val Graveglia, is located 68 metres above sea level (the lowest area), while Santo Stefano d'Aveto is located at 1012 metres above sea level (more than 900 metres higher). The municipalities have been divided according to the SNAI classification criteria in terms of distance from the nearest service provision centre: Borzonasca and Ne are the closest to their reference pole, located near the municipal boundaries of the town of Chiavari (27,335 inhabitants). The areas classified as intermediate are altogether five, and four of them gravitate around the pole of Genoa (Bargagli, Davagna, Lumarzo and Torriglia), while Mezzanego refers to Chiavari. The areas classified as peripheral are characterized by a higher average altitude (Propata in Antola Valley reaches 990 metres above sea level) and, apart from the exceptions of Rezzoaglio (867 inhabitants) and Rovegno (482 inhabitants), have a population resident not exceeding 240 units (Fontanigorda). The municipality of Santo Stefano d'Aveto, the only one classified as ultra-peripheral, is located at the top of the Valley (1012 metres) but has more residents (980). The table also shows that the total number of inhabitants of the inner area is equal to 16,464 while the average population density

Table 4.1 Municipalities of Antola-Tigullio area

Municipality	Inner area classification by SNAI	Inhabitants (2022)	Total area (sq. km)	Population density (pop/sq. km)	Altitude (m)
Borzonasca	Outlying	1798	80.51	22	167
Ne	Outlying	2099	63.52	33	68
Bargagli	Intermediate	2538	16.28	156	341
Davagna	Intermediate	1824	20.53	89	552
Lumarzo	Intermediate	1412	25.51	55	228
Mezzanego	Intermediate	1489	28.65	52	83
Torriglia	Intermediate	2185	60.02	36	769
Fascia	Peripheral	70	11.25	6	900
Fontanigorda	Peripheral	240	16.16	15	819
Gorreto	Peripheral	93	18.88	5	533
Montebruno	Peripheral	210	17.68	12	655
Propata	Peripheral	117	16.93	7	990
Rezzoaglio	Peripheral	867	10.72	8	700
Rondanina	Peripheral	60	12.81	5	981
Rovegno	Peripheral	482	44.09	11	658
Santo Stefano d'Aveto	Ultra –peripheral	980	54.78	18	1012
Inner area Antola-Tigullio		16.464	592.3	33	

Source: Author's elaboration based on PFA (2017)

is 30.60 inhabitants per sq. km (significantly lower than the average value of the Ligurian inner areas of 50 inhabitants per sq. km).

To fully understand the characteristics of the Antola-Tigullio inner area, it is necessary to frame the socio-economic aspect of the territory, its productive fabric and its offer of essential services such as mobility, education, health and digital divide.

The data on the resident population of the area, as reported in the Project Framework Agreement of 2017, has remained almost unchanged from the mid-1980s to the present day. This is due to the balancing effect between the growing suburbanization of the two referring centres of Chiavari and Genoa and the depopulation of the most remote areas.

Other socio-demographic trends of considerable interest for these territories concern ageing of population and old-age index: The over-65 population residing in Antola-Tigullio inner area (30% in 2015) ranks above both the national and regional averages. In addition, the old-age index of this area—that is the ratio between over-65 population and children under 14—is clearly higher than the Italian (157.7) and the Ligurian average (242.7): the Antola-Tigullio area presents an index of 289.3 elderly every 100 children.

In terms of production, the territory is characterized by the presence of a small number of companies and of small size: the sectors most represented in the area are construction (30%) and agriculture, forestry and fishing (21%). The tourism sector in these valleys has profoundly changed in recent years: the typical tourist is attracted today by the countless offers of outdoor activities that this area can offer (trekking, e-bikes, hiking, fishing, etc.). This tourist trend partially replaces the old concept of vacation that characterized citizens of the main neighbouring centres who decided to move to these places, mainly in summer, to escape the heat of the city. Based on this, the tourism sector of these territories has also attracted many foreign tourists in recent years (mainly Dutch, Swiss and German).

Table 4.2 shows the main characteristics of essential services offered in these areas.

After describing Antola-Tigullio inner area from a socio-demographic point of view, an overview of the morphological and geographical characteristics of the territory is needed.

The orographic characteristics of a territory greatly affect its degree of accessibility: inner areas, being mainly mountainous territories, present a conformation that it does not adapt well to scheduled transport, paving the way to new and innovative forms of mobility.

Factors having the greatest impact on the accessibility of a territory are its degree of slope ("steepness") and its altitude above sea level ("altitude").

Considering the physical conformation of the Liguria Region, characterized by a coastal strip set between the sea and the Apennines, the Antola-Tigullio area presents very sloping bands near the coast and the city of Genoa (greater than 35% of acclivity) and less steep areas on the north side of the area.

The altitude factor, as previously mentioned, greatly affects the accessibility of the territory. In the Antola-Tigullio inner area, moving away from the coast, the altitude rises to over 1000 metres in the municipality of Santo Stefano d'Aveto. The innermost areas exceed 800 metres in height and represent most of the territory falling within the boundaries of the inner area designated by SNAI. The municipalities close to the coast and to the two attracting poles of Genova and Chiavari are indeed located between 400 and 800 metres above sea level.

The hilly-mountainous conformation of this specific inner area has represented over time a challenge also in terms of urbanization, due to the difficulties of construction and settlement. Woods and crops in fact mainly cover the area, while the inhabited centres are often small and far from each other. The analysis of land use indicates a prevalence of settlements close to the coast and the centre of Genoa (in particular, the municipalities of Bargagli and Davagna), while inland, only the municipality of Torriglia exceeds 2000 inhabitants.

Through the Geoportal on the official website of the Liguria Region, it is possible to download the map relating to the settlement structure of Antola-Tigullio inner area (Fig. 4.4).

The Geoportal map reproduces the settlement structure of the area, distinguishing areas in different colours, each associated with an abbreviation (explained in the legend):

Table 4.2 Essential services offered in Antola-Tigullio inner area

Type of essential service	Description
Mobility	The resident population in the Antola-Tigullio Valley presents one of the most critical conditions (among the four inner Ligurian areas) in terms of accessibility to the nearest service provision centres: Citizens must travel an average of 50.4 minutes by car (the average of the other internal Ligurian areas is equal to 35.7 minutes) to access the essential services they need. Based on these data, it is clear that the public transport system must be entirely redesigned in order to avoid that more and more people move to live in major centres and the process of depopulation of the territory continues. To date, in fact, public transport has a very limited offer of just 6.8 trips/day for every 1000 inhabitants, against the regional average of 8.7. Another important fact is that in the reference area, no citizen lives within 15 minutes of a railway station or a motorway exit and only a minimum portion of the population is able to reach them within 30 minutes
Education	The education offer consists of the presence of 25 schools in the area, all included between kindergarten and first grade secondary school (no presence of second grade secondary school/high schools), which provide for the presence of 1131 students, divided into 74 classrooms (of which 8 are multi-age classes). The data mentioned show a particularly complex situation that involves aforementioned issues
Health	The municipalities belonging to the Antola-Tigullio refer to two different local health authorities, in Italy known as "Aziende Sanitarie Locali" (ASL): 11 municipalities close to the Genoa pole belong to ASL 3 while 5 municipalities close to Chiavari are part of ASL 4. A very important data for understanding the quality of medical care offered in the area concerns the rate of hospital admissions: This data is an indication of how the medical facilities in these areas are not qualitatively or quantitatively able to meet demand and therefore oblige patients to be admitted to hospitals located in the attractive poles of Chiavari and Genoa (even several kilometres away). Based on the above, the rate of hospitalizations for acute cases (number of hospitalizations/total number of residents of the area) in the Antola-Tigullio area is 10.4% (the city of Genoa is at 10.00%), while if considering only the hospitalization rate for the population over 65, it reaches 20.2% (higher than 18.9% in Genoa).
Internet connection	The provision of infrastructures and technological services in the area represents a key element for the relaunch of inner areas: In fact, technology allows the development of innovative services such as telemedicine, remote education services, home working (increasingly widespread in Italy after the outbreak of the coronavirus pandemic). In the Antola-Tigullio area, only a very low share of population (21.6%) has access to a medium-power internet connection: The Italian population that has access to this connection is 36.7% of the total, a much higher percentage than in Antola-Tigullio inner area. Moreover, even more critically, about a quarter of the population of this area has no access to internet and is totally excluded from the opportunities that this technology offers. With this in mind, SNAI considers it crucial in order to develop these territories and avoid the depopulation of inner areas, to spread the broadband connection and allow internet access to more and more people: The on-call transport service, highly technological-based, can only benefit from this vision

Source: Author's elaboration based on PFA (2017)

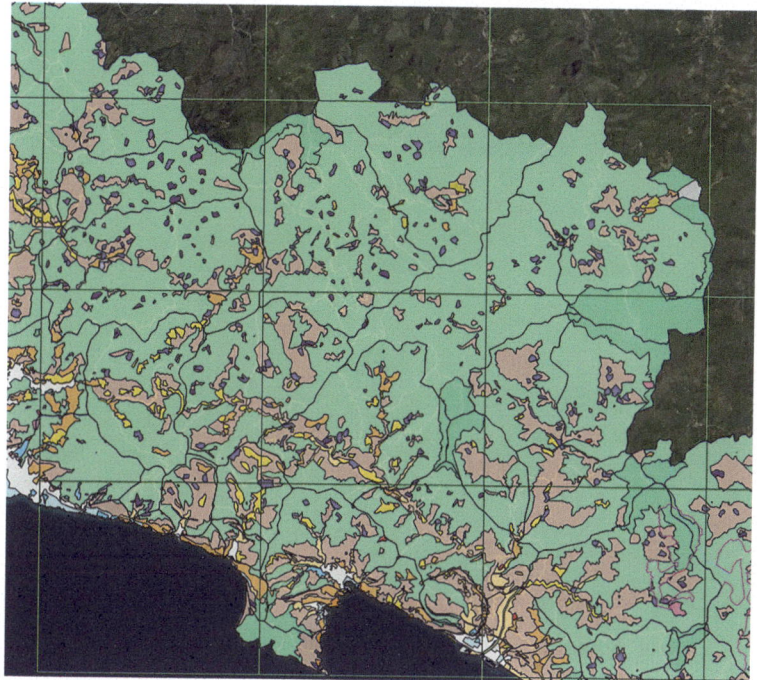

Fig. 4.4 Settlement structure of Antola-Tigullio inner area. Source: *regione.liguria.it*

- ANI—Aree Non Insediate ("Uninhabited Areas") are represented by various shades of green and represent almost the entire territory
- IS—Insediamenti Sparsi ("Scattered Settlements") are represented by various shades of pink
- ID—Insediamenti Diffusi ("Widespread Settlements") are represented by various shades of yellow and orange
- NI—Nuclei Isolati ("Isolated Centres") are represented by various shades of purple.

Figure 4.5 focuses exclusively on the municipality of Fontanigorda. It is possible to observe the subdivision of the residential settlements up close: all around the small town there are uninhabited areas (ANI, in green), gradually followed by scattered settlements (IS, in pink), diffuse settlements (ID, in yellow) and finally an isolated centre (NI, in purple).

As for the road network that affects this area, as showed from Figs. 4.6 and 4.7, it essentially consists of the state road SS45 (Antola Valley) and the provincial road SP586 (Tigullio Valley) up to the village of Rezzoaglio, after which it takes the name of SP654. SS45 state road, built in 1928, connects the cities of Genoa, located on the Ligurian Sea, with Piacenza, in the Po Valley, crossing the entire Apennines on the border between the regions of Liguria and Emilia-Romagna. The "provincial road SP586 of the Aveto Valley" also connects the regions of Liguria and

Fig. 4.5 Focus on Fontanigorda's settlement structure. Source: *regione.liguria.it*

Emilia-Romagna, as well as the state road SP654 that reaches the highest point of the Ligurian Apennines.

These roads, which represent the main access to these hilly-mountainous territories, branch off at various points in minor roads (municipal) to reach the more isolated hamlets. The conformation of a road network of this kind entails the need, for implementing a DRT service, to use vehicles of adequate size to travel along winding and very narrow roads in order to reach even the most remote stops.

After analysing the characteristics of the territory in terms of the provision of services and at the morphological level (studying the steepness, altitude, land use and settlement structure of the area), it is crucial to deepen the study of social and economic dynamics that characterize this inner area.

The first factors taken into consideration concern the territorial dimensions of municipalities, number of residents and population density. These values, shown in Table 4.3, provide a concrete picture of the situation affecting this area.

In terms of territorial size, Rezzoaglio (Tigullio Valley) represents the largest municipality (104.72 sq. km) with a population of less than 900 units and a very low population density (just 8 inhab./sq. km). On the contrary, the municipality of the smallest inland area is Fascia in Antola Valley (11.25 sq. km) with a population of only 70 people and one of the lowest population densities (6 inhab./sq. km) together with Rondanina and Gorreto (5 inhab./sq. km both). For these contexts, an innovative and inclusive transport system is essential to guarantee the population access to essential services and thus counteract the phenomenon of depopulation and abandonment of these areas.

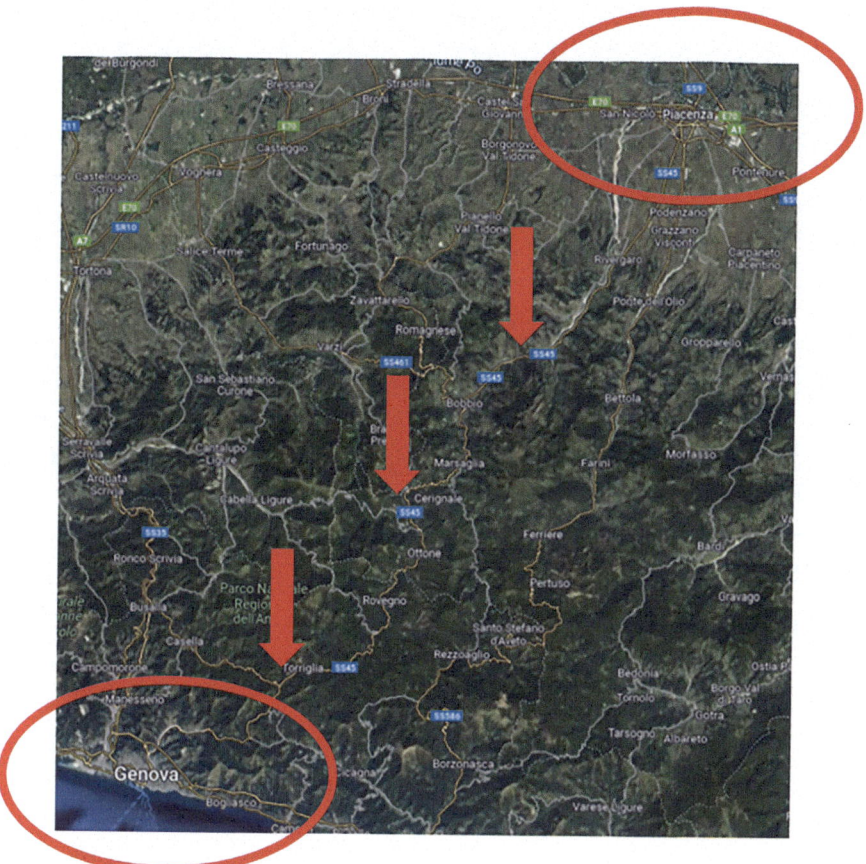

Fig. 4.6 State Road SS45 connecting Genova and Piacenza through Antola Valley. Source: Screenshot taken by the author on *Google Maps*

From Table 4.3, it can be observed that municipalities located near the attractive poles of Genoa and Chiavari are the most populous. For example, Bargagli is immediately located outside the municipal boundaries of Genoa and, despite a very low territorial extension (just 16.28 sq. km), had 2538 residents in 2022, with a population density of 156 inhab./sq. km (the highest data among the municipalities of the inner area under examination). The municipality of Davagna, adjacent to Bargagli, also presents the same characteristics (thanks to its proximity to the city of Genoa, it presents 1824 inhabitants with a land area of 20.53 sq. km).

The Chiavari centre, on the other hand, has a rather high population for the neighbouring municipalities of the inner area such as Ne (2099 inhabitants), Mezzanego (1489 inhabitants) and Borzonasca (1798 inhabitants). While Borzonasca and Ne have a rather large territorial dimension (80.51 sq. km and 63.52 sq. km, respectively) and therefore a low population density (22 and 33 inhab./sq.

Fig. 4.7 SP586 branches off into SP654 at Rezzoaglio, to reach Santo Stefano d'Aveto. Source: Screenshot taken by the author on *Google Maps*

Table 4.3 Main characteristics of Antola-Tigullio municipalities

Municipality	Land area (sq. km)	Resident population (2022)	Population density (inhab./sq. km)	Altitude (m)
Bargagli	16.28	2538	156	341
Borzonasca	80.51	1798	22	167
Davagna	20.53	1824	89	552
Fascia	11.25	70	6	900
Fontanigorda	16.16	240	15	819
Gorreto	18.88	93	5	533
Lumarzo	25.51	1412	55	228
Mezzanego	28.65	1489	52	83
Montebruno	17.68	210	12	655
Ne	63.52	2099	33	68
Propata	16.93	117	7	990
Rezzoaglio	104.72	867	8	700
Rondanina	12.81	60	5	981
Rovegno	44.09	482	11	658
Santo Stefano d'Aveto	54.77	980	18	1012
Torriglia	60.01	2185	36	769

Source: Author's elaboration based on ISTAT data

km, respectively), the municipality of Mezzanego has rather small dimension (28.68 sq. km) and a higher population density (52 inhab./sq. km).

The municipalities of Torriglia and Lumarzo, despite their position being distant from both cities, take advantage of their central position in Antola Valley, which determines a rather high population for both contexts (2185 and 1412 inhab./sq. km, respectively). If Torriglia has large territorial dimensions (60 sq. km) and therefore a low population density (36 inhab./sq. km), Lumarzo has the opposite characteristics (25.51 sq. km and 55 inhab./sq. km). The municipalities of Rezzoaglio and Santo Stefano d'Aveto, located in the deep inland of Antola-Tigullio inner area, present very low population densities (8 and 18 inhab./sq. km, respectively) due to large and sparsely populated territories.

Altitude is a factor inversely proportional to the accessibility of a territory and therefore to its populousness. Excluding the exceptions of Santo Stefano d'Aveto, Torriglia and Rezzoaglio, all the most populous municipalities are located at a very low altitude (the most populous municipality, Bargagli, is located at 341 metre above sea level).

Another relevant element for understanding the socio-demographic dynamics of the area concerns the depopulation trend of these territories. Table 4.4 compares data on the population surveyed in 2011 and 2019 (ISTAT), providing an overview of the population changes of Antola-Tigullio inner area over time.

The analysis of ISTAT data shows a demographic decline that has affected all the municipalities of the inner area in the last decade, with the sole exception of Montebruno that instead recorded a slight increase (+4% residents). The highest

Table 4.4 Depopulation trend of Antola-Tigullio municipalities

Municipality	Legal population (2011)	Resident population (2019)	Population lost	Depopulation %
Bargagli	2810	2613	−197	−7%
Borzonasca	2124	1946	−178	−8%
Davagna	1927	1842	−85	−4%
Fascia	100	65	−35	−35%
Fontanigorda	274	253	−21	−8%
Gorreto	107	85	−22	−21%
Lumarzo	1594	1504	−90	−6%
Mezzanego	1624	1488	−136	−8%
Montebruno	217	225	+8	+4%
Ne	2361	2203	−158	−7%
Propata	161	137	−24	−15%
Rezzoaglio	1080	949	−131	−12%
Rondanina	69	62	−7	−10%
Rovegno	568	506	−62	−11%
Santo Stefano d'Aveto	1217	1097	−160	−13%
Torriglia	2392	2219	−173	−7%

Source: Author's elaboration based on ISTAT data

percentages of depopulation concern the municipalities of the deep hinterland: the highest numbers in the period 2011/2019 were recorded in Antola Valley in the municipalities of Fascia (−35%), Gorreto (−21%) and Propata (−15%). The municipalities of deep hinterland of Tigullio Valley also report significant demographic drops, in particular Rezzoaglio (−12%) and Santo Stefano d'Aveto (−13%). These data show how the great distance from the coast and from the attractive poles that characterizes the aforementioned municipalities inevitably entails conditions of marginalization, social and economic exclusion and therefore, in the absence of targeted support initiatives, a progressive depopulation of these areas.

On the contrary, the municipalities closer to the coast recorded a minor demographic decline in the period considered, also due to the effect of new residents who decided to leave the city looking for a higher life quality and more accessible housing market. The municipality of Davagna, adjacent to Genoa, recorded the smallest decline (−4%) while Borzonasca (−8%), Bargagli (−7%) and Ne (−7%) remained almost stable. The municipalities of Torriglia and Lumarzo, due to their centrality in the Antola Valley, have also maintained their population rather stable in the period 2011–2019 (demographic decline of −7% and −6%, respectively).

The municipality of Montebruno, on the other hand, is surprisingly the only one to have increased its population over the last decade (growth of +4%).

At this point of the analysis, the phenomenon of the depopulation is observed over a very large period: the 2018 report on inner areas published by ASITA (Geomatics Federation in Italy) provides interesting data on the percentage variation of the population of Antola-Tigullio Valleys in the reference period 1951–2017.

The inner area under examination suffered the greatest depopulation in the post-war period (1951–1971) where it recorded a residents' decline of 28.35%. This negative trend has gradually decreased over time (−17.90% in the period 1971–2001) up to a slight increase of +3.61% in 2001–2011. From 2011 to 2017, the depopulation rate started to decrease again (−4.47%).

The greatest percentage variation in population (from −70% to −30%) happened, confirming what aforementioned, in the most remote and difficult-to-access municipalities such as Rezzoaglio, Santo Stefano d'Aveto, Gorreto, Fascia, Propata, Rovegno, Fontanigorda, Rondanina, Montebruno and Propata. The municipalities of Borzonasca and Ne recorded a demographic decline of lesser intensity in the period 1971–2011 (from −30% to 0). Furthermore, the municipalities close to the coast and to the centres of Genoa and Chiavari are characterized by a slight demographic increase (from 0 to 50%).

Another element to take into consideration in terms of socio-demographic dynamics concerns the phenomenon, relevant in recent years, of foreign immigration in the municipalities of the area closest to the coast and to the cities. This phenomenon developed because of foreign families (made up of people of working age and minors) willing to move for work reasons from cities to inner areas: this process helped in compensating the demographic decline and in lowering average age of these municipalities.

ASITA reported also data showing the incidence of foreign immigration in the municipalities of the area: in the decade 2001–2011, the foreign population of

Antola-Tigullio inner area increased by 383.48% (followed by a slight decrease of 1.80% in the following period 2011–2017). According to 2017 data, it should be noted that the phenomenon of foreign immigration particularly affected the municipalities of Borzonasca, Mezzanego, Rovegno (foreign presence between 10% and 15%) and Propata (foreign presence between 5% and 10%). This has no particular incidence in the other municipalities of the area.

It is crucial now to focus on the population that, instead of leaving, decided to stay and live in these areas. Understanding the population composition in terms of groups of age and travel behaviours is essential in order to organize an ad hoc transport service in this context. Table 4.5 shows data relating to the resident population in 2017, the percentage division by age group, the old-age index and average age of each municipality.

The municipalities with the highest percentage of children under 14 are Mezzanego (14%), Bargagli (12%), Ne (12%), Lumarzo (11%), Davagna (11%) and Montebruno (10%). The young people belonging to this age group are characterized by the lack of their own means of transport (as they do not have the necessary age to obtain a driving licence) and therefore their demand for a transport service tailored to their needs could be higher (mainly home-school trips). All the other municipalities of the inner area show a presence of young people under 14 below the 10% threshold: the borderline cases are represented by Gorreto (4%), Rondanina (2%) and above all Fascia (0%), all part of the Antola Valley.

The age group 14–64, relative to the active population, represents the largest share of the inhabitants of almost all the municipalities: people between 14 and

Table 4.5 Groups of age, old-age index and average age of Antola-Tigullio municipalities

Municipality	Resident population (2017)	% 0–14	% 14–64	% > 65	Old-age index	Average age
Bargagli	2697	12	64	24	202.1	46.4
Borzonasca	2089	9	57	33	356.4	51.4
Davagna	1895	11	61	29	263.9	49.1
Fascia	75	0	39	61		66.6
Fontanigorda	266	6	46	49	806.3	58.1
Gorreto	97	4	44	52	1250	64.7
Lumarzo	1502	11	58	31	296.2	49.6
Mezzanego	1544	14	61	25	175.1	45.3
Montebruno	238	10	55	35	350	50.6
Ne	2252	12	62	26	216.5	47.7
Propata	140	6	56	38	588.9	53.8
Rezzoaglio	982	7	49	44	593.1	56.4
Rondanina	64	2	48	50	3200	59.7
Rovegno	533	8	57	36	475	53.6
Santo Stefano d'Aveto	1122	7	59	35	505.2	53.1
Torriglia	2297	9	61	30	316.2	50.5

Source: Author's elaboration based on ISTAT data

64 years show the greatest demand for mobility to reach work, study and leisure places. The municipalities near the coasts and the cities (Bargagli, Davagna, Mezzanego and Ne) all have a percentage share greater than 60%, as does Torriglia (61%).

Out of 16 municipalities, only 4 show a lower share of the active population (14–64) than the over 65. These municipalities are Fascia (active population 39% compared to 61% of over 65), Fontanigorda (46% compared to 49%), Gorreto (44% compared to 52%) and Rondanina (50% compared to 48%).

Furthermore, the aforementioned municipalities are also those with a more advanced average age of the population (66.6 years, 58.1 years, 67.7 years and 59.7 years, respectively).

Another very important element concerns the so-called "old-age index". It measures the number of elderly (over 65) present in a population of every 100 young people (under 14). This index provides information on the degree of ageing of a population and values over 100 indicate the presence of a majority of elderly compared to young people. For this reason, the municipality of Fascia has an old-age index that cannot be measured, as the presence of young people under 14 is zero.

The municipalities with the highest share of population over 65 are also those with a higher old-age index: the highest value in absolute corresponds to Rondanina (3200), followed by Gorreto (1250) and Fontanigorda (806.3). On the contrary, all the municipalities adjacent to the cities or with a central position in the valley have a rather low old-age index (but in any case, higher than 100). Bargagli (202.1), Davagna (263.9), Lumarzo (296.2) and above all, Mezzanego (175.1) represent the lowest values.

To understand travel behaviours of population residing in these areas, it is important to observe the number of private vehicles available in each municipality (both cars and motorbikes). This data provide significant information about the degree of car/motorbike dependence of citizens and therefore their willingness to use any new public transport service.

Table 4.6 shows the number of cars and motorbikes present in each municipality and the number of cars per 1000 inhabitants in order to provide indications on the dependency rate of the population on the use of private vehicles.

The municipalities with the highest number of cars (data as of 2016) are also those with the highest resident population: Bargagli, Torriglia, Ne, Borzonasca, Davagna and Lumarzo. On the contrary, the lowest numbers are found in the municipalities with the smallest population such as Fascia, Gorreto, Propata and Rondanina.

The values relating to the availability of motorbikes reflect the data on cars: the greatest number of motorbikes is concentrated in the most populous municipalities.

The data about the number of cars per 1000 inhabitants indicate that Lumarzo has the highest value (699) and therefore the highest car-dependence of its citizens. Lumarzo is followed by the municipalities of Torriglia (644), Rondanina (641) and Bargagli (600).

Table 4.6 Number of cars and motorbikes of Antola-Tigullio municipalities

Municipality	Number of cars (2016)	Number of motorbikes (2016)	Total	Cars per 1000 inhab.
Bargagli	1617	583	2200	600
Borzonasca	1133	364	1497	542
Davagna	1123	463	1586	593
Fascia	37	11	48	493
Fontanigorda	145	51	196	545
Gorreto	54	12	66	557
Lumarzo	1050	335	1385	699
Mezzanego	925	322	1247	599
Montebruno	131	43	174	550
Ne	1453	533	1986	645
Propata	72	22	94	514
Rezzoaglio	530	111	641	540
Rondanina	41	16	57	641
Rovegno	332	114	446	623
Santo Stefano d'Aveto	664	160	824	592
Torriglia	1480	455	1935	644

Source: Author's elaboration based on ISTAT data

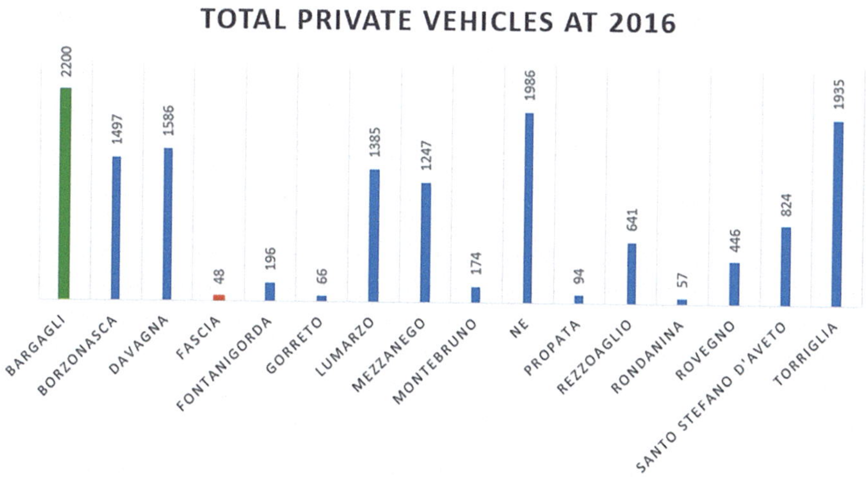

Fig. 4.8 Total private vehicles (2016) of Antola-Tigullio municipalities. Source: Author's elaboration based on ISTAT data

From the aforementioned analysis, the situation of strong car-dependence of population emerges: the residents of these municipalities prefer using private vehicles because of the scarcity of public transport service and the convenience and flexibility of cars. Figures 4.8 and 4.9 graphically illustrate the total number of private vehicles (2016) and the data relating to the number of cars per 1000 inhabitants, respectively.

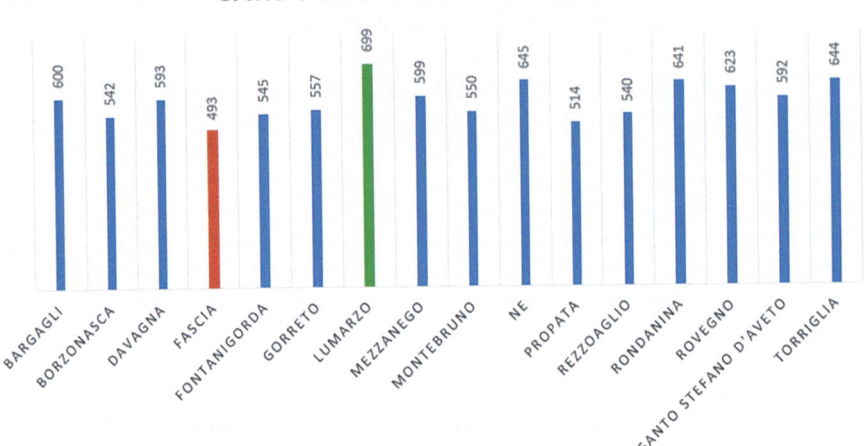

Fig. 4.9 Cars per 1000 inhabitants (2016) of Antola-Tigullio municipalities. Source: Author's elaboration based on ISTAT data

Current Transport Offer in Antola-Tigullio Valleys

Before one can understand the need of implementing a DRT service, one must examine the traditional public transport offer existing in Antola and Tigullio Valleys.

The municipalities of Antola-Tigullio inner area, based on the subdivision carried out by AMT (transit provider of the entire urban and metropolitan area of Genoa), belong to two different service lines: the municipalities of Antola Valley belong to "Valbisagno and Alta Valtrebbia" ("Group M") while territories of Tigullio Valley belong to the "Tigullio Centrale" line. The municipality of Lumarzo is the only one to be served by two different lines: the aforementioned "Tigullio Centrale" and the one called "Golfo Paradiso", which includes the areas closer to the coast.

To the "Valbisagno and Alta Valtrebbia" service line belong the following municipalities:

- Genova (not part of Antola-Tigullio inner area project)
- Bargagli
- Davagna
- Fascia
- Fontanigorda
- Gorreto
- Montebruno
- Montoggio (not part of Antola-Tigullio inner area project)
- Propata
- Rondanina
- Rovegno
- Torriglia

Table 4.7 "Valbisagno and Alta Valtrebbia" service line

Service line no.	Service line name
725	Torriglia—Bargagli—Genova Brignole
829	Buffalora—Torriglia—Laccio—Scoffera
726	Torriglia—Scoffera—Davagna—Prato
824	Traso Ponte—Traso Centro—Traso Alto—S. Alberto—Bargagli
910	Bromia—Laccio—Torriglia
925	Montebruno—Rovegno—Gorreto—Ottone
926	Torriglia—Casoni—Fontanigorda—Casanova—Carchelli—Rovegno
927	Torriglia—Propata—Rondanina
928	Local lines Torriglia
929	Local lines Fascia
930	Local lines Fontanigorda—Rovegno—Montebruno
931	Local lines Gorreto

Source: Author's elaboration

The "Valbisagno and Alta Valtrebbia" service consists of the lines contained in Table 4.7 (only lines serving Antola-Tigullio inner area municipalities are reported).

The two traditional public transport lines 831 (Prato-Marsiglia) and 832 (Prato-Terrusso-Cisiano) have been replaced by an on-call transport service active since 13 June 2022. The "Chiama il bus" service, active from Monday to Saturday (06.00–19.30) serves the localities of Bargagli, La Presa, Preli, Viganego, Terrusso, Cavassolo, Maggiolo, Calvari, Marsiglia and Capenardo, all adjacent to the Genoa pole. The seat on the bus can be booked through the call centre or through a dedicated app.

To the "Tigullio Centrale" service line belong the following municipalities:

- Bargagli
- Borzonasca
- Carasco (not part of Antola-Tigullio inner area project)
- Chiavari (not part of Antola-Tigullio inner area project)
- Cicagna (not part of Antola-Tigullio inner area project)
- Cogorno (not part of Antola-Tigullio inner area project)
- Coreglia Ligure (not part of Antola-Tigullio inner area project)
- Favale di Malvaro (not part of Antola-Tigullio inner area project)
- Genova (not part of Antola-Tigullio inner area project)
- Lavagna (not part of Antola-Tigullio inner area project)
- Leivi (not part of Antola-Tigullio inner area project)
- Lorsica (not part of Antola-Tigullio inner area project)
- Lumarzo
- Mezzanego

- Moconesi (not part of Antola-Tigullio inner area project)
- Ne
- Neirone (not part of Antola-Tigullio inner area project)
- Orero (not part of Antola-Tigullio inner area project)
- Rezzoaglio
- S. Colombano Certenoli (not part of Antola-Tigullio inner area project)
- S. Stefano d'Aveto

The "Tigullio Centrale" service consists of the lines contained in Table 4.8, classified by AMT according to the municipality of reference (only lines serving Antola-Tigullio inner area municipalities are reported).

Table 4.8 "Tigullio Centrale" service line

Municipality of reference	Service line no.	Service line name
Bargagli	715	Chiavari FS—Gattorna—Ferriere—Genova
	725	Genova Brignole—Bargagli—Torriglia—Ottone
	824	Traso ponte—Traso centro—Traso alto—S. Alberto—Bargagli
Borzonasca	712	Chiavari—Borzonasca
	711	Chiavari—Borzonasca—Rezzoaglio—S.Stefano d'Aveto
	917	Borzonasca—Levaggi—Belpiano—Acero
	916	Borzonasca—Caregli
	915	Borzonasca—Borzone
	812	Borzonasca—Belvedere
	826	Carasco—Montemoggio—Borzonasca
Lumarzo	715	Chiavari FS—Gattorna—Ferriere—Genova
Mezzanego	712	Chiavari—Borzonasca
	711	Chiavari—Borzonasca—Rezzoaglio—S.Stefano d'Aveto
	826	Carasco—Borgonovo—Montemoggio—Borzonasca
	812	Borzonasca—Belvedere
	814	Chiavari FS—Villagrande di Cichero—Ferreccio
Ne	703	Chiavari FS—Lavagna Osp.—S.Salvatore—Conscenti
Rezzoaglio	911	Rezzoaglio—Alpepiana—Vicomezzano—Vicosoprano
	912	Rezzoaglio—Sbarbari
	711	Chiavari—S.Stefano d'Aveto
	913	Rezzoaglio—Casaleggio
	994	Rezzoaglio—Torrini—Pareto—Ascona
Santo Stefano d'Aveto	711	Chiavari—S.Stefano d'Aveto
	994	Rezzoaglio—Torrini—Pareto—Ascona

Source: Author's elaboration

Many lines of traditional public transport service in the municipality of Ne have been replaced by an on-call transport service; it will be described in detail in the next chapter. To the "Golfo Paradiso" service line belong the following municipalities:

- Genova (not part of Antola-Tigullio inner area project)
- Avegno (not part of Antola-Tigullio inner area project)
- Bogliasco (not part of Antola-Tigullio inner area project)
- Cicagna (not part of Antola-Tigullio inner area project)
- Lumarzo
- Moconesi (not part of Antola-Tigullio inner area project)
- Neirone (not part of Antola-Tigullio inner area project)
- Pieve (not part of Antola-Tigullio inner area project)
- Recco (not part of Antola-Tigullio inner area project)
- Sori (not part of Antola-Tigullio inner area project)
- Tribogna (not part of Antola-Tigullio inner area project)
- Uscio (not part of Antola-Tigullio inner area project)

The only line of the "Golfo Paradiso" service to serve the municipality of Lumarzo is number 870 "Ferriere—Lumarzo—Colle Caprile".

Chapter 5
Results of the Survey of Mayors Conducted in Antola-Tigullio Inner Area

5.1 Hypothesis of DRT Services in Antola-Tigullio Inner Area

This chapter describes the process that led to the experimental implementation of on-call transport in some selected municipalities of the Antola-Tigullio Valleys. It uses a survey, conducted through a questionnaire addressed to all the municipalities' mayors, as a first element useful for planning a DRT service as close as possible to the real needs of users.

The decision was made to exclusively interview the mayors of the communities concerned and exclude the general public in order to get the most impartial and reliable information about the area. The mayors perfectly suited this goal since they represent the whole community they are a part of. The questionnaire method was used to get responses that were consistent and similar as much as possible. Since the DRT service proposed integrates/replaces existing lines of FT and for which it was believed to already have a solid socio-demographic and mobility database, a GIS accessibility study of the region was not performed for the objectives of this research.

The answers to this inquiry allowed me to address the following research question, investigating elements considered crucial in literature, as described in paragraph 6.1, for a successful implementation of the service:

RQ4: "What are the perceptions of mayors of inner areas concerning the characteristics of the territory and the residents' travel behaviour with regard to the implementation of DRT services?"

After understanding, also through the valuable suggestions provided by mayors, the strengths and weaknesses of public transport offer and the mobility needs of residents, three pilot areas were identified where to initially test the new DRT service. A hypothesis of DRT service was then associated with each selected pilot area.

T. Pavanini, *Rural Demand Responsive Transport*, SpringerBriefs in Operations Management, https://doi.org/10.1007/978-3-031-91395-2_5

This resulted in the definition of a new on-call transport service to supplement existing FT. Furthermore, in order to test the feasibility of the hypothesized solutions, a field inspection was conducted by CIELI researchers (University of Genoa) and AMT managers. In conclusion, a workshop was organized with all stakeholders involved to discuss the findings of this study.

For the purposes of this work, only one of the three hypothesized services is taken into consideration and analysed, as it is the first that found concrete application: the pilot started on 14 February 2022.

5.1.1   Study of Population Travel Behaviours: The Questionnaire

The questionnaire submitted to the mayors of the municipalities of Antola-Tigullio inner area consists of three sections: in the introduction, the purpose of the survey and the fundamental characteristics of the new on-call transport service are explained to the mayors. Then, in the central part of the questionnaire, seven open-ended questions, aimed at fully understanding the demand characteristics and the users' travel behaviours, are expressed. Finally, the mayors are asked to indicate the name of their municipality and to leave their contact details (email and telephone).

The first question aims to collect information about three different characteristics of each municipality: population, accessibility and services.

The mayors are initially asked for some demographic characteristics such as the number of inhabitants, the average age, the average income and the general economic and social conditions in which the population lives. Demographic information is crucial in order to implement an ad hoc DRT service, as reported by Jain et al. (2017). The accessibility of the territory is useful to understand through which infrastructures the municipality can be reached (both outgoing and incoming) and with which means of transport (whether private or public). As revealed by Laws et al. (2009), the territory accessibility represents a key factor in the implementation of DRT services: their study, conducted on DRT cases in England and Wales, showed that one of the main drivers incentivizing local authorities to invest in this technology is represented by the will to increase the area accessibility for citizens.

The mayors are also required to list the essential services present within the municipality and, if necessary, for those absent, to which other municipality the citizens turn to meet their needs.

The second question is aimed at understanding which categories of users encounter the greatest degree of difficulty in terms of mobility within the municipal area and towards the outside. Mayors are also asked to provide information on the time slots less served by traditional public transport: this information allows users to be targeted in order to plan an ad hoc DRT service. The importance of user segmentation based on specific characteristics for the success of DRT services has been already demonstrated in academic literature (Vij et al., 2020; Woolf & Joubert, 2013; Nyga et al., 2020).

The third question still intends to obtain information about the existing FT offer: in particular, the mayors are asked for the interventions they consider a priority for improving the transport service (better routes, stops and services). The Mishra et al. (2022) study states how the integration of the new DRT service with the existing traditional transport is strictly necessary to provide users with a complete transport offer.

The fourth question asks mayors to indicate from which origins to which destinations the new DRT service could be more useful: this depends on the presence or absence of essential services in the relevant municipality. Based on this information, it is possible to outline the correct ad hoc DRT service models described in Sect. 2.5 of this elaborate as identified in literature (Nelson et al., 2004; Papanikolaou et al., 2017; Mageean & Nelson, 2003).

The next question aims to understand the possible availability of local private entities to collaborate with the public transport service provider, as is already the case in various parts of the world (Jain et al., 2017): in this way, public-private synergies can be realized and contribute to improving the general offer of the transport service in such areas. The sixth question collects information about the possibility of using passenger transport to also offer different services to the residents of these mountain areas: for example, public transport vehicles can also transport medicines (Hirsch & Fredericks, 2001; Schumacher, 2020; BBC, 2023), food and mail to the more remote and difficult to access locations.

The seventh question, like the second, aims at obtaining an in-depth targeting of users. Specifically, the objective is to attract more tourists to these areas, through the introduction of changes to the current transport system. Almost all the municipalities of the Antola-Tigullio inner area present a purely summer type of tourism linked to sports such as trekking, hiking and biking. The routes and stops that means of transport serve must be rethought with a view to favouring tourists in reaching the main desired destinations. The importance of combining on-call transport with tourism purposes has already been demonstrated by Matsuhita et al. (2022), who tested the "Misato Aiai Taxi" service in the town of Aizumisato (Japanese prefecture of Fukushima).

The eighth question investigates the propensity of residents of the area to use technologies such as smartphones, computers, tablets, applications, etc. This information is useful for creating a DRT transport service with user-friendly booking methods easily accessible even by older users, who show a natural propensity to use technology after appropriate training courses (Burlando & Cusano, 2014).

The last question aims to collect further indications and suggestions from the mayors of the area: in particular, it tries to study the best possible integration between traditional public transport and flexible on-call services.

Except for Torriglia's mayor, who did not respond, all of the mayors sent their responses to the questionnaire (response rate: 94%).

5.1.2 Analysis of Questionnaire Responses

The answers to the first question are interesting above all in relation to the accessibility of the territory and the availability of essential services. Most of the mayors pointed out how access to their municipalities takes place through municipal, provincial and state roads, allowing connection with larger centres both by car and FT vehicles. Some mayors also complained about the lack of maintenance carried out on the road surface, which makes the viability dangerous (Propata, Fontanigorda and Rondanina).

From the observation of the results relating to the presence of essential services (e.g. post office, pharmacy, supermarket, bank, etc.), a rather heterogeneous picture emerges. The 20% of the municipalities (3) have all the essential services within their territory: Bargagli, Borzonasca and Mezzanego—located near the Genoa and Chiavari poles—have a large population and provide all the essential services that citizens require. An opposite situation prevails in four municipalities of the Antola-Tigullio inner area (26.67%), whose mayors declared they do not have any essential service on their municipal territory and that therefore their citizens are forced to travel to the neighbouring major centres. This data is understandable for the municipalities of Fascia, Gorreto and Propata (where a butcher also sells foodstuffs) due to the difficult accessibility of their territories and the low resident population. On the contrary, the case of Ne is surprising: though more populated, the population must go to the neighbouring Conscienti centre to get access to essential services.

The municipality of Fontanigorda presents essential services on its territory but complains about reduced hours and few days of opening: in addition to the lack of a butcher, the clinic opens only once a week, the pharmacy and the post office on alternate days while the bank service is available only one afternoon a week. Two municipalities in the area (Montebruno and Rezzoaglio) present only some essential services while the residents have to travel to neighbouring municipalities for others. In Montebruno, there are medical services, pharmacies, post offices, banks and police but not school services. In Rezzoaglio, on the other hand, there are clinics with specialists present once a month, post office, bank and primary school, but for all other services, residents must go to Bedonia (PR), Chiavari, Genoa or Santo Stefano d'Aveto.

It should be noted that five municipalities in the area did not answer this specific question or in any case answered in an irrelevant way.

Figure 5.1 shows responses in terms of essential services for a better and immediate understanding.

The analysis of the results of the second question makes it possible to target the categories of users who, in the opinion of mayors, find it more difficult to access mobility within the municipality borders and towards the main neighbouring centres.

All the mayors included the elderly among the people who find most obstacles to mobility. This fact is not surprising, as the elderly, who represent the prevalent component of the population of these areas, very often do not have a private vehicle to

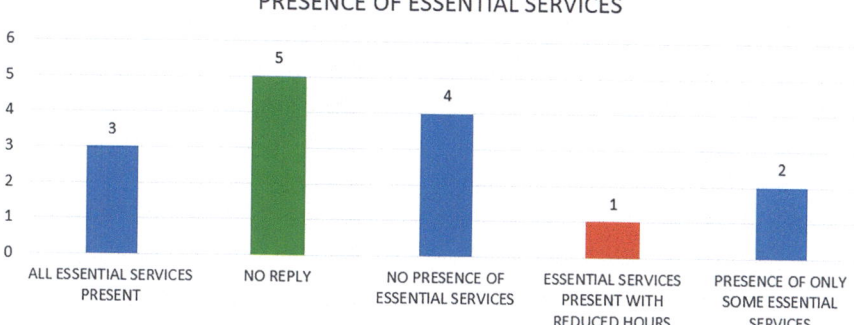

Fig. 5.1 Presence of essential services. Source: Author's elaboration

move and, due to a poor public transport service, consequently face the risks of isolation. Many mayors pointed out that the elderly residing in these areas need an efficient transport service especially during the morning hours (10.00–12.00), those less served by FT, in order to reach the neighbouring municipalities for services such as the bank, post office and pharmacy. Mayors indicated even evening time slots as critical: scheduled line service is not available and therefore, the elderly and other categories of users that do not own a car experience difficulties with mobility.

Ten mayors indicated the category of young people as a target of users who find difficulties in access to mobility: this is because many children under 14 are not of the age to obtain a moped licence and therefore cannot reach the major interchange centres (and under 18 cannot drive a car). Time slots indicated as critical are referred to morning: high school students often have to make very long trips to reach the cities of Bedonia in Emilia-Romagna (especially young people residing in Rezzoaglio) or Chiavari and Genoa. Very often, students have to get up very early in the morning and return home in the late afternoon due to the absence of a more suitable transport service for school hours.

Only three municipalities (20%) indicated the category of workers: Borzonasca, Davagna and Ne. The workers residing in these areas, due to the lack of an effective traditional transport service, prefer to use their own car. The mayor of Davagna pointed out that even the category of workers of foreign origin residing in the municipality faces numerous mobility problems, often not owning a private vehicle.

The mayors of Borzonasca and Fontanigorda also indicated the category of tourists: in particular, hikers, trekkers and mountain bikers encounter mobility problems associated with moving from the poles of Genoa/Chiavari to the start of trails and back.

Figure 5.2 illustrates categories of users afflicted by mobility problems, providing a better and immediate understanding.

The third question asked mayors to indicate the priority interventions required, taking into account the existing public transport offer. 40% of them expressed a desire to strengthen the existing FT system by introducing new routes (Davagna, Mezzanego, Ne, Rezzoaglio, Rovegno) and stops (Rondanina).

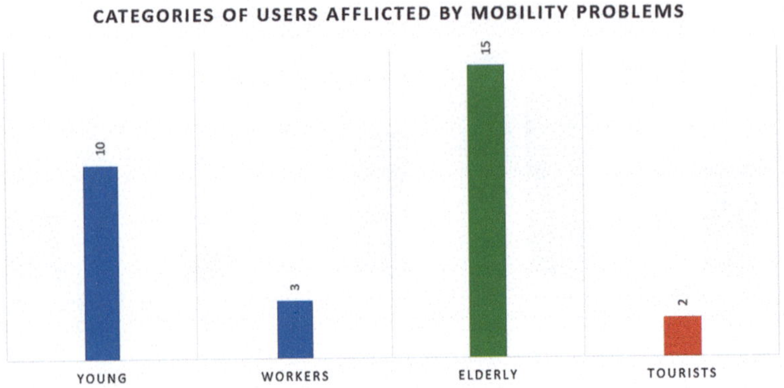

CATEGORIES OF USERS AFFLICTED BY MOBILITY PROBLEMS

Fig. 5.2 Categories of users reported by mayors as afflicted by mobility problems

The mayors of four municipalities requested a better connection of the existing public transport with the hamlets scattered within the municipal boundaries: together with the municipalities of Rezzoaglio and Rondanina, the mayor of Gorreto asked for connections from the more isolated hamlets to nearby major centres such as Gorreto capital, Rovegno, Montebruno and Ottone.

The mayors of Borzonasca, Ne and Mezzanego requested the implementation of an on-call transport service. The municipality of Borzonasca considered crucial the presence of a DRT service exploiting Borzonasca capital as an interchange node between FT and on-call service. The municipality of Ne requested the introduction of an on-call service capable of transporting people with motor disabilities, while Mezzanego considered the DRT service necessary for hilly areas within its municipality.

Three municipalities (20%) wanted to define an interchange node between scheduled and on-call transport services in correspondence with their capital. In addition to Borzonasca, the mayors of Propata and Bargagli also expressed this need.

Three mayors (20%) also expressed the need for a better school transport: the municipality of Montebruno considered it important to introduce a much faster school bus service to the Genoa hub, Propata required the implementation of a dedicated students line to reach the primary school in Torriglia and Santo Stefano d'Aveto asked for a dedicated line for students going to Bedonia (PR).

Finally, two mayors expressed different requests: the municipality of Fascia requested a transport service (without specifying whether FT or DRT) whose trips have the same timetables as the main essential services. The municipality of Rondanina expressed its willingness to adapt public transport timetables to the needs of residents and tourists (and therefore the introduction of a DRT service is considered appropriate).

Figure 5.3 illustrates priority interventions as reported by mayors.

In response to the fourth question, all the mayors expressed their opinion by indicating as a priority the connections from the more peripheral hamlets to the larger centres offering essential services. The mayor of Rondanina pointed out the

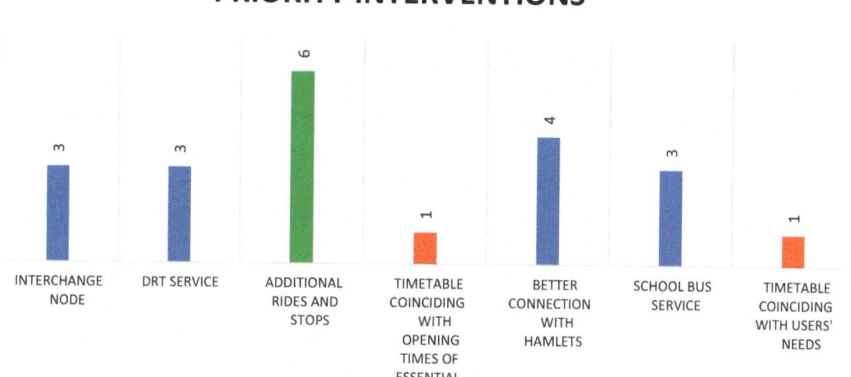

Fig. 5.3 Priority interventions

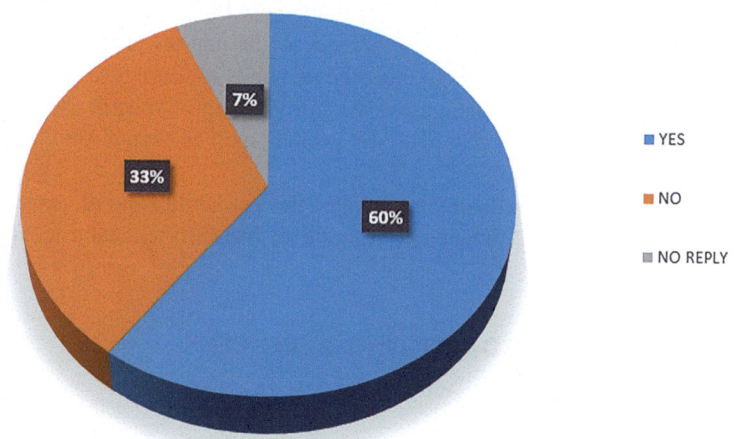

Fig. 5.4 Possibility of public-private collaboration in mayors' perception

lack of a connection to the lake of Brugneto, a well-known tourist and sports resort: he considered it necessary to introduce a transport service dedicated to this locality at least in the spring-summer period.

The fifth question, relating to possible collaborations between the public transport provider and private entities, showed 60% of positive opinions: the mayors replied that collaborative relationships between public and private companies are already active, in particular for school transport and car with driver. Figure 5.4 shows the mayors' perception of the possibility of establishing partnerships between public and private entities (%).

The sixth question of the questionnaire investigated the need to transport medicines, mail and foodstuffs to these areas, in addition to people, and asked mayors to indicate locations most in need. Most of the answers indicated the need for the transport of medicinal products: this is the result of the generalized ageing process that affected these areas in the last decades. The majority of requests (60%) concerned the transport of foodstuffs, while others required the delivery of mail. In addition, mayors could propose further transport solutions: the municipality of Gorreto required the transport of newspapers, while Rondanina requested transport services for submitting complaints to the police. The mayor of Santo Stefano d'Aveto supported the need to use vehicles allowing the transport of bicycles for tourists interested in mountain biking in the area. The municipality of Montebruno, on the other hand, did not detect any need to transport other goods besides people. Figure 5.5 summarizes mayors' requests in terms of transporting additional products in the DRT vehicles.

Question number 7 asked mayors to indicate possible changes to be made to the transport service (new routes, stops, shuttles, etc.) in order to attract more tourists in those areas.

Except Bargagli, Fontanigorda and Montebruno, the mayors of all the other municipalities (80%) responded positively to the possible extension of the transport service to the more touristic locations. Almost all the mayors promoted the implementation of new rides and stops to meet the demands of tourists, especially in the summer period. Propata also suggested the introduction of specific stops to serve the main tourist destinations also in winter period (such as Monte Antola and Casa del Romano).

Question 8 of the questionnaire investigated the availability and propensity of the residents of the inner area under examination to use info-telematic technologies

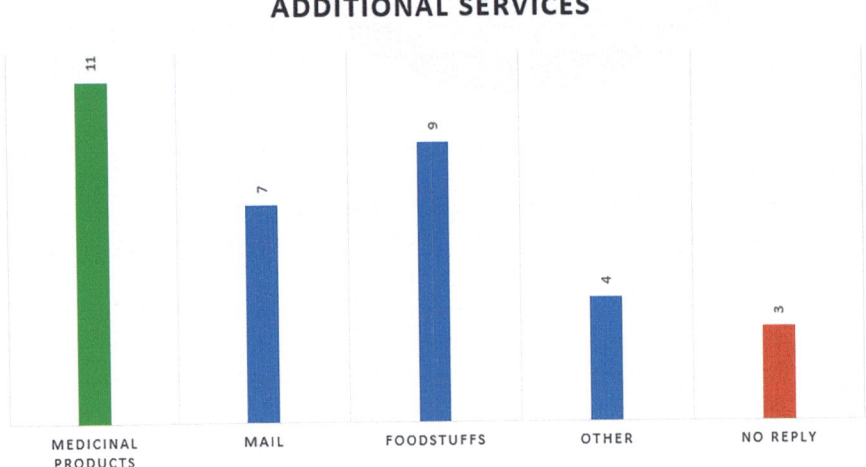

Fig. 5.5 Additional services

(smartphones, tablets, computers, applications, etc.), which are essential for booking an on-demand transport service.

The responses of the mayors showed a heterogeneous picture. Many of them indicated at the same time more options due to a propensity for technology, which naturally varies according to the age of residents. Most of the responses indicated a low propensity for technology: not surprising given the average advanced age of the population of these areas. A small number of mayors stated a medium level of availability to use technology and only a few considered it high (where the elderly population is not predominant). These data provided important information in relation to the most suitable booking method for the new DRT service to be implemented: it appears logical, with a rather low propensity to use technology in this area, to avoid providing the use of the website or a dedicated app as the only possible booking methods. On the contrary, a booking method based on telephone calls appears more suitable for the context.

Figure 5.6 illustrates the level of population propensity to the use of technology.

As regards question 9, asking the mayors for further suggestions about the services to be implemented in the area, no additional details emerged. They only reiterated the importance of strengthening the existing FT and the utility of a good quality DRT service tailored to the local mobility demand.

From the questionnaire submitted to the mayors of Antola-Tigullio inner area, important information was collected for the design and implementation of a DRT transport service: key factors learnt from this inquiry are stated below.

What emerged from the survey can be summarized as a general attention and priority given to the elderly population: the mayors requested better connections from the most remote hamlets to the main centres offering essential services, booking methods suitable for an older age group (such as the possibility to book up to the

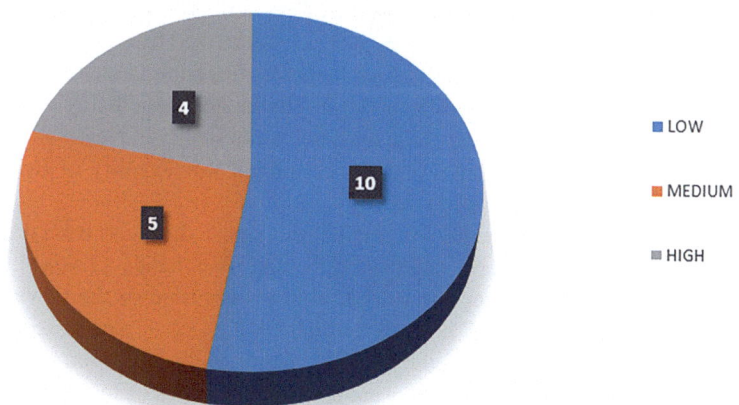

Fig. 5.6 Technology propensity level of population

day before by phone) and times of the rides corresponding to the needs of this user range (mainly in the morning).

The other category of users strongly afflicted by the mobility problems typical of these mountain areas were students: many mayors highlighted how young people, not having their own vehicle, have to spend many hours on board public transport vehicles to reach the nearest schools (even 2 hours in some cases). In order to prevent students from having to get up very early in the morning and return home late in the afternoon, the mayors are calling for the introduction of faster service lines with fewer stops.

For the category of workers, the introduction of an on-call service is less of a priority as people in this range almost all have their own vehicle, which they would hardly abandon in favour of public transport.

Some mayors also expressed the need to create a DRT transport service capable of meeting the needs of tourists who visit these areas for activities such as hiking, trekking and mountain biking. For this category of users, the introduction of seasonal service lines (mainly in summer) or on holidays is considered necessary.

Finally, almost all of the mayors interviewed considered it useful to transport other goods on the same vehicles intended for passengers—particularly medicinal products, foodstuffs and mail.

5.1.3   DRT Service Hypothesis

The survey conducted made it possible to collect a series of data useful for understanding the mobility dynamics of the categories of users residing in the inner area, to analyse the strengths and weaknesses of the current public transport offer and to design a new, more efficient and flexible DRT service.

At this stage, taking into consideration the mayors' recommendations regarding the users' needs and the areas' socio-demographics, three pilot municipalities were selected to design three distinct DRT services. These are profoundly different areas, which, based on the SNAI classification, were characterized by a different distance from the Genoa and Chiavari poles (main characteristics summarized in Table 5.1).

The analysis led to the selection of the intermediate municipality of Bargagli (small size but highly populated), of the outlying municipality of Ne (larger size and population over 2000 inhabitants) and the peripheral municipality of Rezzoaglio (large size and very low population density).

It is decided, for each of the pilot municipalities selected, to start from the recognition of the current transport offer in different time bands (7.00–11.00 and 15.00–19.00) to be able to capture the variability over time of the service, with particular attention to the "soft" hours for which DRT is actually proposed.

The next step is then represented by the actual proposal of a new on-call service. For each area, the main characteristics are specified in terms of locations reached, timetables and booking systems.

Based on the requests expressed by the mayors in the surveys, the main objectives that each of the proposed solutions offer are reported.

Table 5.1 Main characteristics of Bargagli, Ne and Rezzoaglio

	Bargagli	Ne	Rezzoaglio
SNAI classification	Intermediate	Outlying	Peripheral
Self-containment	Low	Low	High
Type of settlement	Widespread settlement	Scattered settlement	Isolated centres
Population	2613 inhab.	2252 inhab.	982 inhab.
Population density	Highest in the inner area (165.95 inhab./sq. km)	Low (34.95 inhab./sq. km)	Very low (9.21 inhab./sq. km)
Territorial size	Small (16.28 sq. km)	Medium (64.1 sq. km)	Very large (104.7 sq. km)
Prevailing demographic groups	Under 14 14–64 years	Under 14 14–64 years	Over 65 14–64 years
Motorization rate	High	Very high	Very low

Bargagli

The existing public transport offer in the municipality of Bargagli consisted of the presence of five different service lines (15, 24, 25, 26, 27) which, in particular in the soft hours, did not sufficiently connect the fractions of the municipality with the capital of Bargagli. In the questionnaire, the mayor of Bargagli expressed the need for a better connection between the hamlets of Viganego, Terrusso, Cisiano and Maxena with the capital Bargagli in late afternoon-evening hours. For this reason, a very rigid service model with fixed routes and fixed stops (score of flexibility 1) was assumed in integration to the FT and active from 05.00 to 20.00, Monday to Saturday.

Taking into account the requests of the mayors, a proposed DRT service intends to connect the fractions farthest from the SS45 with Bargagli capital and with the pole of Genoa to take advantage of long-distance transport lines. In addition, the service reduces the difficulty of accessing mobility for the elderly, students and the disabled.

Ne

The offer of traditional public transport existing in the municipality of Ne consists of three different lines (3, 31, 34). The mayor's intention was to strengthen the existing offer through a better hourly coverage of the service and the introduction of an on-call service for the elderly category. The request of the mayor of Ne was a better connection with the town of Conscienti, home to the main services of the municipality. The centre of Conscienti was thus assumed the ideal interchange hub for passengers residing in the surrounding hamlets.

Thus, an on-call service was proposed, which provides for transit through the hamlets of Conscenti, Camminata, Frisolino, Pian di Fieno, Chiesanuova,

Casedogana, Pontori, Ne and Castagnola, active from Monday to Saturday, 07.00 to 19.00. The proposed service presented flexible timetables and routes, with fixed stops (score of flexibility 3). Furthermore, this can replace traditional transport line 34.

The introduction of a DRT transport service of this type entails better service coverage in the late morning and early afternoon time slots, as requested by the mayor. The service allows students to reach school complexes faster.

Rezzoaglio: S. Stefano D'aveto

The third proposal concerns the largest municipalities of the inner area, namely Rezzoaglio and Santo Stefano d'Aveto. The existing FT lines in the area were 11, 111, 112 and 113.

The municipal areas of Rezzoaglio and Santo Stefano d'Aveto are among the largest in Liguria. Priority requests of mayors were essentially two: on the one hand, to improve the fraction-capital connections for the elderly in the mid-morning time slots; on the other, to guarantee a school service to students allowing them to reach the schools of Chiavari and Bedonia (PR) faster than today. The service hypothesis consisted in the connection of the areas of Gavadi, Amborzasco and Ascona to Santo Stefano d'Aveto and of Villanoce, Alpepiana and Parazzuolo to Rezzoaglio: this service, active 05.00–20.00 from Monday to Saturday, was planned with score of flexibility 1 (fixed stops and a predetermined itinerary).

An on-call transport service such as the one proposed above provides a greater territorial extension, greater hourly coverage, increased accessibility for the elderly and disabled. This service is also provided through call centre booking the day before.

5.1.4   The Inspection

In conclusion, to testing feasibility of above proposals, on 15 October 2021, a joint inspection was carried out between researchers of CIELI (University of Genoa) and AMT managers in charge of DRT project for the Antola-Tigullio inner area.

This inspection concerned exclusively the routes of DRT transport service assumed for Bargagli, to which was also added the inspection of the Davagna area. This because AMT managers decided that this service area would have been one of the first to test the trial (started on 13 June 2022). Localities of Prato, Viganego, Terrusso, Cisiano, Bargagli, Sant'Alberto di Bargagli, Davagna, Capenardo and Cavassolo were then inspected.

The route of the inspection was chosen by AMT managers to validate on the field the sections of the route that the future on-call transport vehicles must travel in these areas. The meeting place was at the main square of the municipality of Bargagli where it was decided to continue towards the terminus of the urban line 13 in Prato/

Genoa to begin the inspection. The Prato/Pian Martello urban stop represents the terminus of the urban line 13, which from Prato (outskirts of Genoa) transports passengers to Caricamento (centre of Genoa) and indeed, it constitutes an interchange point for the extra-urban lines arriving from Antola Valley.

From the Prato/Pian Martello stop, the route started towards Bargagli capital on state road 45: the first detour made it possible to reach the remote hamlets of Viganego (isolated centre), Terrusso (larger centre) and Cisiano (isolated centre), which must be connected to the SS45 and to the capital Bargagli through DRT service.

Given the limited width of the road travelled to reach these hamlets, it is considered necessary to use a rather small vehicle capable of manoeuvring easily.

Route continued towards S. Alberto di Bargagli: the road necessary to reach this hamlet, the SP82, is different from the previous one in width (it is wider and able to accommodate vehicles of standard size) and in residential fabric (presence of numerous multifamily houses).

Planned itinerary continued towards the hamlets of the municipality of Davagna, west of Bargagli. The road to reach Davagna capital from Bargagli is the SS45 (which changes its name to "SP62" after Bargagli) that turns off into the SP14 at Scoffera deviation: here, after passing the hamlets of Scoffera, Moranego, Sella and Villa Mezzana, vehicles can reach Davagna, the capital. This stretch of road is characterized by rather large dimensions, suitable for standard buses, and the presence of some shops especially near the junction of Scoffera and the capital itself.

After Davagna, the inspection route continued up to the hamlet of Capenardo, rather remote compared to the main road SP14 and reachable by a particularly narrow and impervious road: it seems logical to use a vehicle of small size and able to drive on steep slopes (presence of snow and ice is typical in winter).

The route then descended towards the city of Genoa through the hamlets of Calvari, Mareggia and Maggiolo (SP14) until the centre of Cavassolo, where road becomes steeper up to the city.

5.1.5 SWOT Analysis

Before the actual experimentation of the on-call service in selected areas of Antola-Tigullio Valley, a workshop was organized between the main project actors: mayors of the municipalities, AMT, University of Genoa and representatives of ANCI (National Association of Italian Municipalities). Some interesting reflections have arisen from this dialogue.

First, the mayor of Davagna requested the introduction of an additional stop at the hamlet of Paravagna (located along the road that connects the capitals of Bargagli and Davagna): DRT service seems to be the ideal solution to meet the mobility needs of residents of this fraction.

The preliminary meeting also revealed the need, in a subsequent phase to the experimentation and if results would have demonstrated its validity, to extend DRT

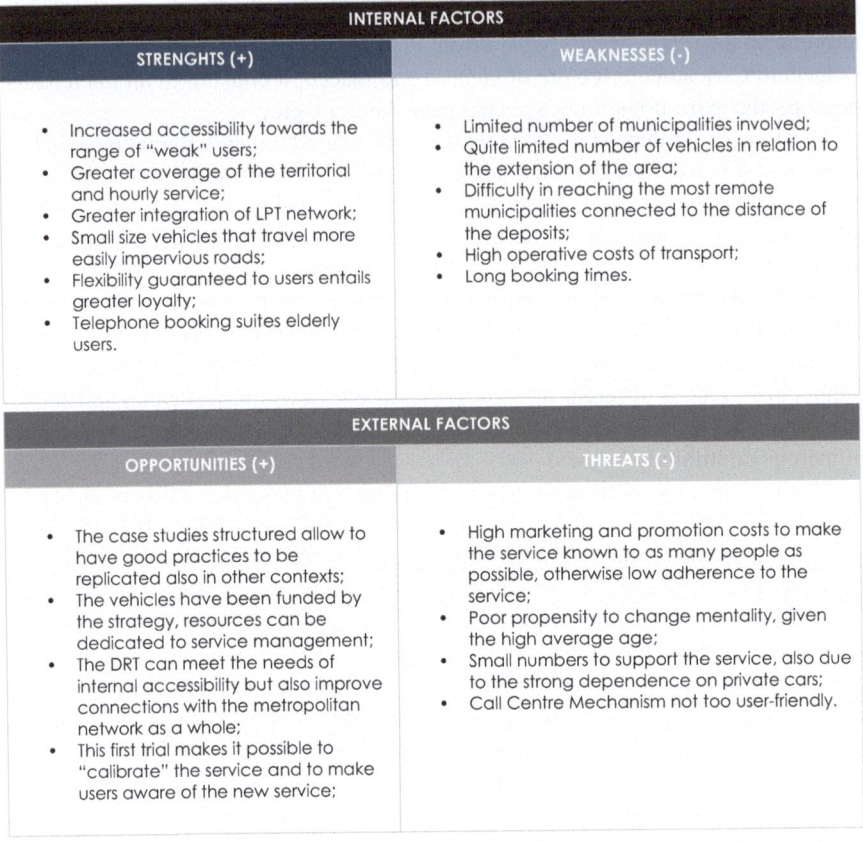

Fig. 5.7 SWOT analysis of proposed DRT services

service to other municipalities in the inner area not taken into consideration in the pilot phase.

Another important issue that emerged during the debate concerned the modelling of an ad hoc DRT service for the tourist category. The satisfaction of the mobility needs of this user group represented one of the main requests for most of the mayors. In the experimentation phase of the project, however, due to the season in which the pilot would have been launched (winter-spring), it was considered a priority to favour the categories of elderly, students and workers residing in the area. A specific DRT service for summer outdoor activities practitioners was postponed to the subsequent stages of implementation of the service.

At the end of the report, it was considered crucial to summarize the main perspectives of the service hypotheses reported using the SWOT analysis tool: to this end, Fig. 5.7 shows the main strengths and weaknesses of the proposed DRT services, as well as their opportunities and threats.

Chapter 6
Ex-Post Evaluation of Val Graveglia DRT Pilot Results

6.1 Ex-Post Evaluation of Val Graveglia DRT Pilot Results

AMT has recently launched several on-call transport services ("Chiama il bus") in various territories, some belonging to the inner areas identified by SNAI, such as the one in Val Graveglia (Ne municipality), in the Bargagli and Davagna area and in the territory of Borzonasca (Fig. 6.1), and others external to this initiative, such as the DRT service operating in Val Brevenna, Recco, Casarza Ligure and Cogorno.

As regards the services of inner areas, this section only reports the results of the DRT service of Ne (Val Graveglia), representing the first trial launched by AMT in the area (14 February 2022): data provided by AMT for the reference period 14/02/2022–19/07/2022 are studied to carry out an initial analysis of the service performance. For the additional DRT services operating in Bargagli and Davagna area (started on 13 June 2022) and in Borzonasca (started on 3 October 2022), it is still too early to analyse the performance data.

This chapter, based on similar studies conducted by Coutinho et al. (2020) and Yen et al. (2023), intends to address the following research question:

RQ3: "How did the DRT service perform when tested in Antola-Tigullio Valley?"

6.1.1 Method for the Classification and Framing of Rural DRT Services

Before going into the analysis of the performance of the DRT service tested in the case study of the Antola and Tigullio valleys, it is necessary to introduce the framework used to frame the characteristics of this type of service. This is essential in order to guarantee maximum replicability of this model, even in contexts other than the one covered by this study.

T. Pavanini, *Rural Demand Responsive Transport*, SpringerBriefs in Operations Management, https://doi.org/10.1007/978-3-031-91395-2_6

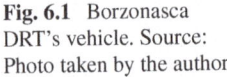

Fig. 6.1 Borzonasca
DRT's vehicle. Source:
Photo taken by the author

In the study carried out by the undersigned (Pavanini, 2023) and published in the
Revista de Estudios Andaluces (REA), an analysis was carried out with the aim of
providing a general overview of the rural DRT services present in Italy, framing and
reporting their main technical characteristics. The aim of this process is to clearly
understand the current layout and the state of technological maturity of DRT in a
given country and to be able to classify these services, highlighting their strengths
and weaknesses.

After a careful study of the Project Framework Agreements of all the Italian
regions and a keyword web search, 35 rural DRT cases active on the Italian territory
since 2010 were identified. It was decided to start the search at the beginning of the
last decade in order to compare the DRT services within the same technological
context.

Once the total number of DRT services had been identified, they were classified
using a specific framework that considered the area of operation, the name, the
booking method, the operator, the time of activity, the cost, the service model and
the presence of a smartphone application.

In addition to these categories, a flexibility score was assigned to each rural DRT service, with values ranging from 1 ("Predetermined route with bookable fixed stops") to 6 ("Many-to-many"), depending on the degree of flexibility adopted by the PTA (see Sect. 2.5 of this book).

Table 6.1 below illustrates this framework model, which can be exported to any context.

6.1.2   Classification of "Ne – Val Graveglia" Rural DRT Service

The on-call transport service in Ne has been active since 14 February 2022 and presents the characteristics proposed in the previous chapter by the report *"Introduction of on-call transport service for the Antola-Tigullio area"* (Table 6.2):

Figure 6.2 indicates the Val Graveglia DRT service area, showing the impervious nature of the served context and the proximity to the Chiavari pole.

Based on the data provided by AMT (related to the period of reference 14/02/2022–19/07/2022), it is possible, as mentioned, to carry out an initial analysis of the operational performance of the service. The seven parameters below are examined one at a time, each with a brief description and a clarifying graph.

Table 6.1 DRT evaluation method

Service Name
Service Area
Service Manager
Period of service activity
Service time slot
Booking system
Cost of service
Route
Stops
App availability
Score of flexibility

Source: Pavanini (2023)

Table 6.2 Ne DRT service characteristics

Service name	"Chiama il bus"
Service area	Conscienti, Caminata, Frisolino, Piandifieno, Botasi, Chiesanuova, Casedogana, Pontori, campo di ne, Iscioli, Castagnola, Zerli, Statale, Reppia and Arzeno
Service manager	AMT
Period of service activity	Pilot, from 14 February 2022
Service time slot	Since 22 June from Mon to Sat h. 06.00–19.30
Booking system	Via app "SERVIZI A CHIAMATA" or telephone (up to 12.00 the day before the ride) It is also possible to use "Chiama il bus" by contacting the driver directly at the Conscenti terminus; each request will be fulfilled compatibly with the reservations previously received.
Cost of service	Same as fixed transport
Route	Flexible
Stops	Fixed, same as FT
App availability	Yes
Score of flexibility	3

Source: Author's elaboration

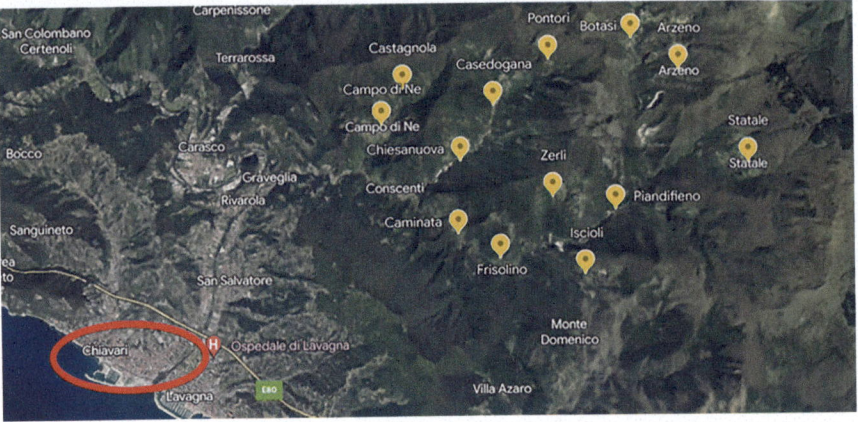

Fig. 6.2 Val Graveglia DRT service area

Booking Method

Database relating to booked rides distinguishes the origin of the booking (call centre or app) for each day of service activity. The data relating to the period considered (total 499 bookings) show a total use of the telephone booking method (100%) confirming the greater usability of this tool especially for the elderly category.

Ride Duration

Data related to the duration of rides for each day of service activity include the average duration of each individual trip (min), the average length (km) and the number of trips carried out.

The average duration and length of trips in the entire reference period are 64 minutes and 24.44 km, respectively. These data show the complexity of operating a traditional transport service in rural or mountainous territories. The average duration of the rides (exceeding 1 hour) and the almost 25 km covered each trip testify the need to find an alternative to FT. Traditional transport indeed is usually characterized by very shorter trips in terms of time and km travelled.

The data show that vehicles completed 369 trips, covering a total of 9388.46 km.

Figure 6.3 illustrates the number of total trial days where a specific number of trips per day were made. As can be seen from the graph, the number of daily trips with higher frequencies is 1 (23 total days), 2 (25) and 3 (34). These results, at the moment, do not require the PTA to purchase additional vehicles for the execution of the service.

Stops Frequency of Use

The database relating to stop usage frequency considers the count of passengers boarding and disembarking from the vehicle at each single stop every day. In addition, the data relating to the frequency with which each stop is used daily by passengers (both for getting on and off) and the number of passages of the vehicle for that specific stop are reported.

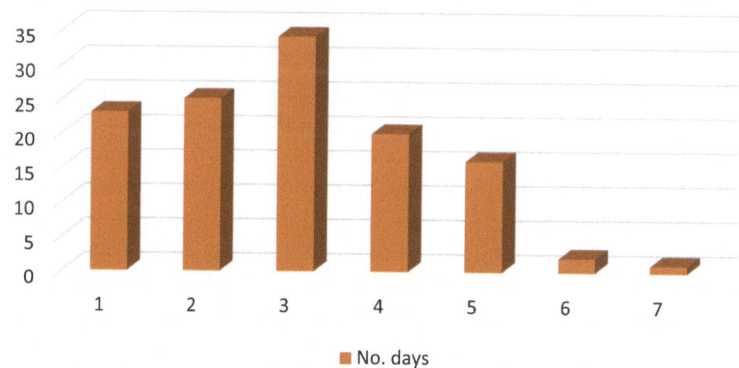

Fig. 6.3 Number of days with X trips/day. Source: Author's elaboration based on AMT data (14/02/2022–19/07/2022)

Vehicles stopped at bus stops 948 times in total, boarding 1114 people (total stop usage frequency).

Figure 6.4 illustrates the frequency of use of each individual stop: "Conscenti Capolinea" is largely the most used stop in the reference period (395 times), amply justifying the mayor's indications to set this location as a strategic interchange point for the whole area.

Booking Requests

The data relating to booking requests received at the call centre (with 100% of bookings made by telephone) are, in the database provided by AMT, divided for each single day of service activity considering the processing status of the requests and the number of seats booked.

Each booking request can thus be classified as "satisfied" if processed in the system by TDC and accepted by users, "not satisfied" if inserted into the system but not accepted by users and "not elaborated" if not even inserted into the system.

The overall total shows 526 requests received at the call centre during the reference period, with 540 seats booked: this data indicates that the vast majority of requests were made to reserve a single seat on board.

Figure 6.5 shows that 89% of booking requests received at the call centre were correctly processed and accepted by users ("satisfied"). In a residual way, 6% of requests were processed in the system but not accepted by users ("not satisfied") and 5% were not even elaborated (trip request not operable).

Figure 6.6 shows that three bookings/day was the highest booking frequency (22 service days), followed by one booking/day and two bookings/day (18 both). These data demonstrate how the call centre, in this scenario, can be efficient even with a few operators given the low number of bookings/day received.

Ride Requests

Data relating to travel requests allow to investigate the number of bookings received at the call centre for each day of service and to classify them based on whether they were carried out in advance or on the same day of the ride.

This information is useful for understanding the needs of the population: people who book a ride in advance, typically the categories of elderly, students and workers, have time to plan their regular trips. In most of the cases, trips booked on the same day pertain to people who use on-demand transport on an occasional basis. Figure 6.7 indicates that 387 reservations were received at the call centre in advance of the day of ride (presumably from the categories of users stated above), while 112 were made on the same day, complicating the service planning process for transit provider.

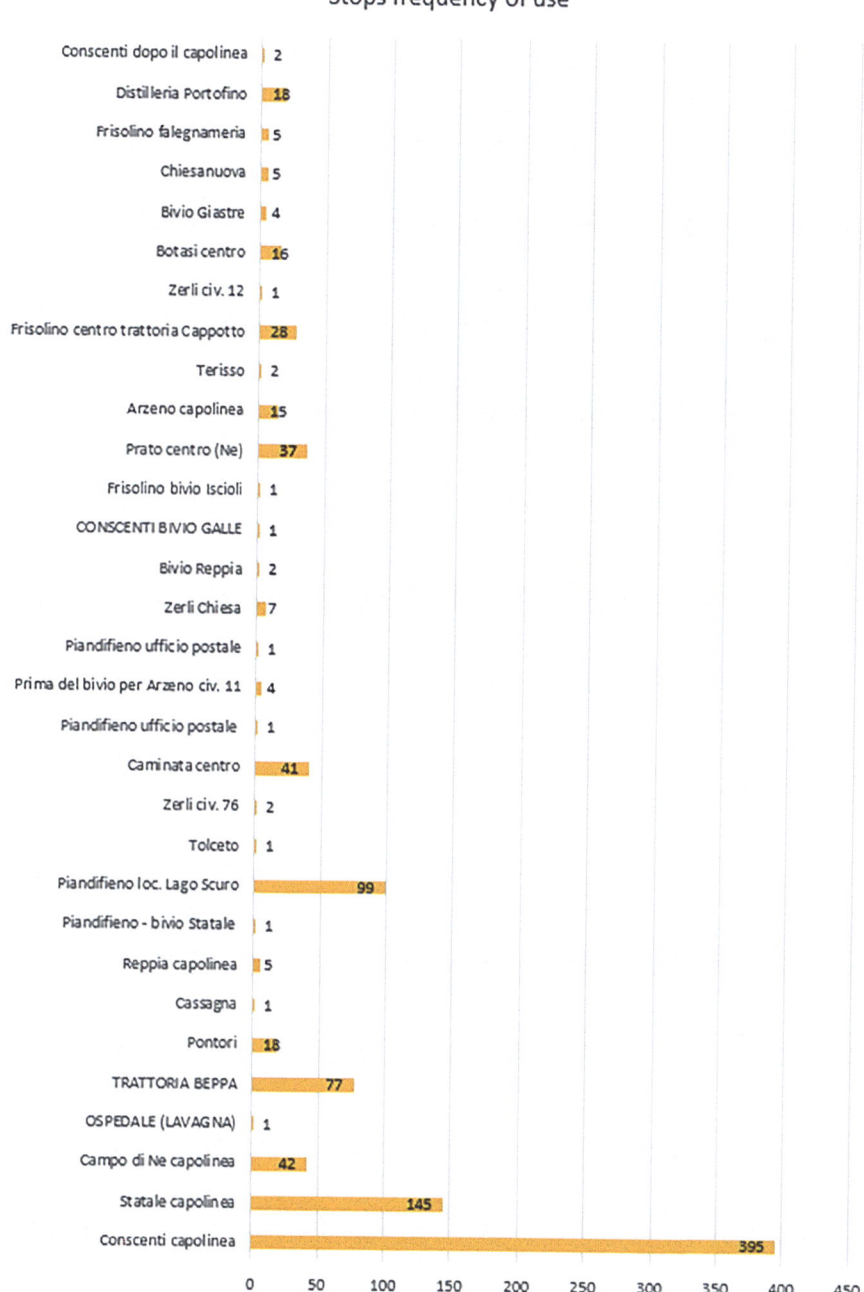

Fig. 6.4 Stops frequency of use. Source: Author's elaboration based on AMT data (14/02/2022–19/07/2022)

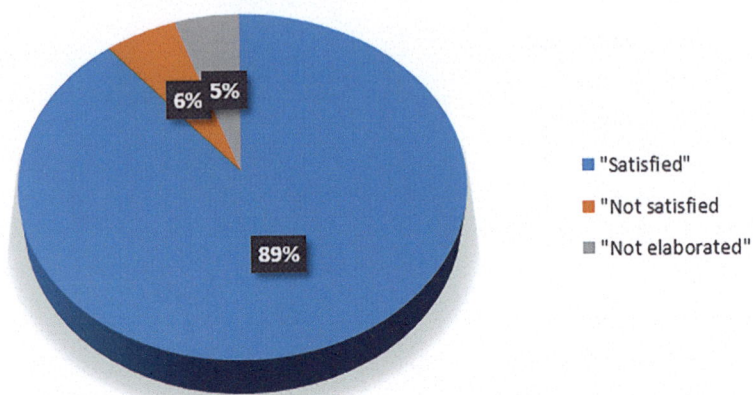

Fig. 6.5 Category of requests. Source: Author's elaboration based on AMT data (14/02/2022–19/07/2022)

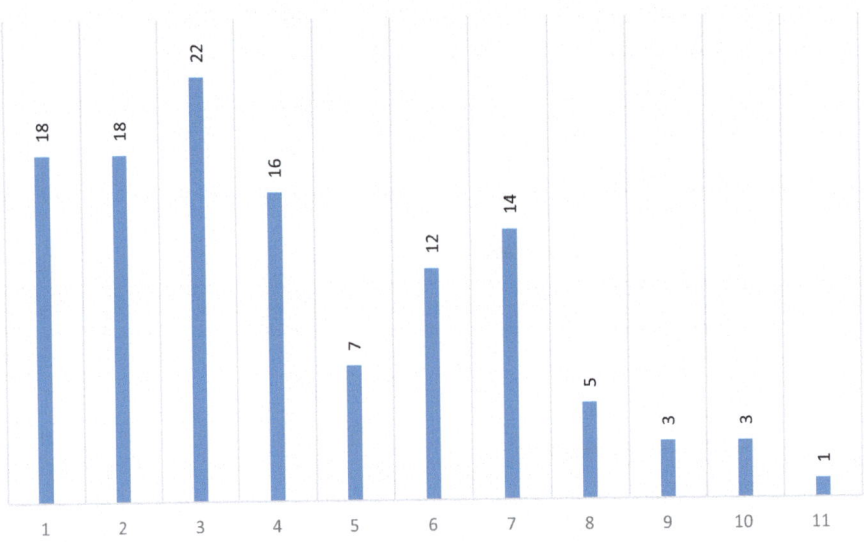

Fig. 6.6 Quantity of requests/day. Source: Author's elaboration based on AMT data (14/02/2022–19/07/2022)

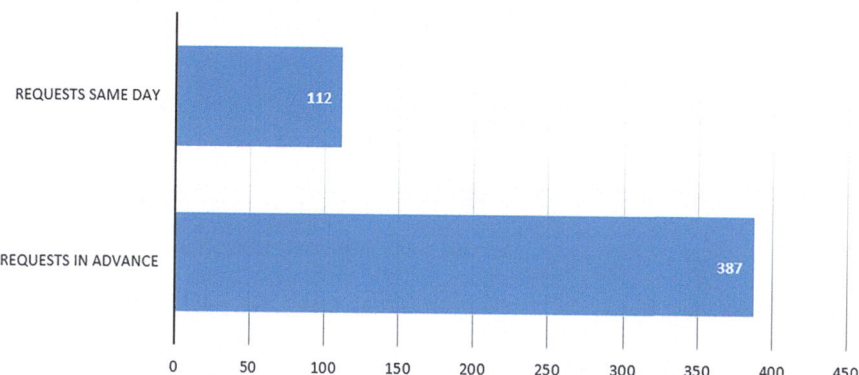

Fig. 6.7 Typologies of requests. Source: Author's elaboration based on AMT data (14/02/2022–19/07/2022)

Service Time

To obtain information about the percentage of use of the vehicles, the availability of the service (number of hours in which the DRT service is operational during the day) is compared with the actual use of the vehicles on that given day.

In addition, daily service time was further divided into service time with passengers on board and total service time (both rides with passengers and rides with no passengers on board to pick them up or to enter/leave the depot). Database shows that service time with passengers on board was 176.19 hours, just over half of the total service time (336.57 hours). Furthermore, total availability of service was equal to 1552.4 hours with a service utilization rate of 21.7%: such a low utilization rate compared to the total availability of service was due to the AMT decision to create an extended daily service time window (11.40 hours per day until 21 June, then 13.50 hours/day). A time window of this width offers an excellent service to users, as it provides maximum flexibility and, at the same time, it entails a considerable cost for the transport company. In future, based on the data available, the transport provider can reason on the creation of a service time window exclusively tailored to the time slots of highest demand for the DRT service. Figure 6.8 illustrates the above.

Users Transported

Finally, the data relating to users transported in the period considered indicate 514 passengers in total.

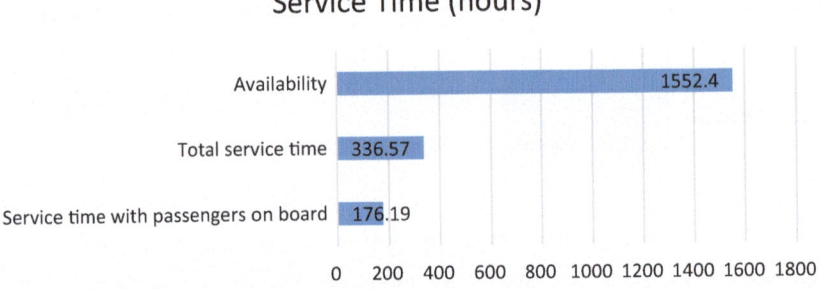

Fig. 6.8 Service time. Source: Author's elaboration based on AMT data (14/02/2022–19/07/2022)

6.1.3 Concluding Remarks

Based on the data provided by the local PTA, it was therefore possible to analyse the passengers' travel behaviours and obtain important information about the main performances of the new on-call transport service implemented in Val Graveglia (Ne).

The resulting information framework, observing the operating period 14/02/2022–19/07/2022, is that of a DRT service much appreciated by the residents of this area and capable of welcoming 514 passengers since its launch. As expected, given the average age of the population residing in Val Graveglia, one of the key elements of the service was not used at all: people decided to book all their rides through the call centre tool, never using the app made available by AMT. Booking via smartphone application represents an important element for the PTAs because it allows them to save personnel costs at the call centre and to manage bookings more efficiently: the fact that no passenger has used this possibility shows decision-makers the great future potential of this service. In this regard, it is necessary to increase the technological knowledge of users and dedicated training courses, organized at some easily accessible points in Val Graveglia, represent a valid solution: as reported by Burlando and Cusano (2014), the elderly of this generation show a natural propensity to use technology that only needs to be stimulated.

The data analysis also indicates that the Conscenti Capolinea stop is the one most used by passengers to get on and off the vehicles of the DRT service. The centre of Conscenti owns the main essential services that meet the needs of all residents of the municipal area of Ne and its surroundings. As per the mayor's request, the report proposed a DRT service that uses the Conscenti centre as an interchange node between the surrounding hamlets and the fixed transport line connecting the municipality of Ne with the poles of the coast (Chiavari, Lavagna, Sestri Levante, etc.). The data show how the questionnaire submitted to the mayors helped in correctly implementing an ad hoc transport service for citizens.

Almost all the bookings received at the call centre (89%) were processed correctly and, after negotiation, accepted by users. Given the low number of daily reservations made (3 represents the highest frequency), the data justify the presence of only a few operators at the call centre, allowing the PTA to save on personnel costs, waiting for a total shift, as mentioned, towards the use of app, which would eliminate this cost entirely.

The data relating to the bookings received in advance of the day of ride show the travel behaviours of passengers: 387 bookings were made in advance and are probably attributable to categories of students, the elderly and workers who have decided to abandon the use of the private vehicle to take advantage of this new service. The relationship between these categories of users and DRT is win-win: on the one hand, passengers can plan their trips in time and on the other, by booking trips in advance, the PTA can better organize the rides to make them as efficient as possible. The 112 reservations made on the same day of the ride can, in any case, be attributed to a good number of people who decide to use the DRT service on an occasional basis. These people have hardly already decided to entrust their trips exclusively to public transport, leaving out car which, as König and Grippenkoven (2020) report, still represents one of the main barriers to the adoption of DRT. Finally, as mentioned, such a low rate of effective use of vehicles (21%) is indicative of an excessively large time window aimed at favouring users by granting them maximum service flexibility. On the basis of the data obtained from this trial, and from others to come, the PTA, in order to save money and energy, can tighten this service window calibrating it perfectly on the travel needs of users.

Chapter 7
Conclusions, Research Limitations and Future Agenda

7.1 Conclusions, Research Limitations and Future Agenda

7.1.1 Statement of Contributions

Starting from the assumption that the characteristics of DRT in rural areas have received little attention in the academic literature, this work seeks to make multiple contributions to the research in this area by examining the global context and analysing the efficiency of a specific Italian case study. The study employed three different methodologies to achieve these goals: international academic literature analysis; distribution of a survey to the mayors of the municipalities of two Ligurian inner areas; and data analysis of DRT pilot results in such regions.

Thus, this work first contributes to the enrichment of the scarce literature on the subject; second, the responses of the mayors of the inner areas involved in the DRT project, along with the analysis of the trial performance, provide further information on the strengths and weaknesses of this tool and its future potential.

7.1.2 Concluding Remarks and Future Agenda

As stated in the introduction of this work, several critical issues in this era call for deep reflections by policymakers regarding the transportation system, both in urban (congestion of roads, climate-changing emissions, noise pollution, most of public space occupied by cars, minimal space for pedestrians, heat islands, etc.) and rural contexts (economic marginalization of the population, social isolation, depopulation of entire territories due to scarce and inefficient transport services and a preference for car use). On-call transport, as observed throughout this work, seems to be one of the potentially valid solutions to the problems affecting mobility today.

T. Pavanini, *Rural Demand Responsive Transport*, SpringerBriefs in Operations
Management, https://doi.org/10.1007/978-3-031-91395-2_7

This work deliberately left out critical issues of the transport system in cities, already widely covered in the academic literature, to focus on the DRT service in contexts characterized by low transport demand and scarce population density, still understudied by researchers. The first two chapters of this work provide useful indications for understanding the need of developing an alternative form of public transport to FT in order to meet mobility needs of population residing in areas difficult to access, sparsely populated and many kilometres away from the main inhabited centres.

Currently, citizens of these areas are forced to use their own cars or motorcycles for daily travel, due to a scarce public transport offer that fails to satisfy their needs. This happens because of the characteristics of these territories that make the provision of traditional transport service economically unsustainable.

In order to encourage people to abandon the use of private means of transport in favour of a new form of mobility—with clear environmental and economic benefits—it is first of all necessary for the PTAs to conduct an in-depth analysis of the transport demand to understand population segmentation and travel attitudes. This allows us to create an on-call transport offer operating in the most requested time slots and addressed to specific user targets.

The analysis of literature review made it possible to study various cases of application of this technology and to understand the reasons behind the numerous failures (RQ1): it was found that although technological progress is strongly linked to the diffusion of the DRT service, greater technological complexity corresponds to high costs for the PTAs. The will of policymakers to offer citizens the greatest possible service flexibility ("many-to-many" model) collides with the economic unsustainability that this entails for the PTAs. Literature also showed that, in the service-planning phase, an in-depth analysis of transport demand turns out to be a key element to avoid providing excessively onerous services. Furthermore, marketing activity through digital and analogue channels also proves to be of primary importance to allow even the elderly, who typically do not use social networks and represent the predominant share of the population in these territories, to get to know the initiative.

The study of the costs associated with on-call transport provided interesting indications: the main cost item is represented by the cost of labour, followed at a distance by the purchase cost of the transport service. Both of these cost items, based on a careful analysis of the transport demand, can be decreased over time by the PTAs: the cost of personnel, as regards the staff of TDC, can be eliminated by encouraging the use of app as a booking tool (removing the need for physical operators at the call centre). Furthermore, in the future, the development of driverless vehicles will be able to eliminate this cost item, which represents the main obstacle to the diffusion of the DRT service.

The rationalization of service can instead be obtained through a better adaptation of the time windows of service operation to the travel behaviours of residents. The analysis of data relating to users' travel attitudes is crucial in correcting this element and thus reducing costs.

Regarding the political vision and the concrete measures adopted by the public entity to address the critical issues of inner areas, Chap. 4 of this work provides some important answers. The Italian government has demonstrated its willingness to remedy to social and economic isolation affecting residents of inner national areas through the adoption, in 2014, of the "National Strategy for Inner Areas". This political measure, part of a wider European initiative, had the objective of halting the depopulation process and creating value for these areas. From the political plan, concrete actions to be implemented in the context emerge, such as a rationalization of health facilities (merger of many small clinics into a few large ones capable of providing basic assistance services, increase the diffusion of first aid tools in the area, develop telemedicine, etc.), a better scholastic offer (better distribution of institutions, more stable contracts for teachers, carrying out school activities related to local culture and traditions, etc.) and a renewed offer of mobility (investments in current transport infrastructure and experimentation with innovative forms of mobility).

RQ2, useful for understanding the main characteristics of the territory and of population in which the DRT service trial takes place, is answered in Chap. 5 of the work. The reference context shows a typical scenario for inner areas: majority of the elderly population, generally low digital maturity, difficulty in accessing the more inland municipalities and greater road connections for the municipalities closest to the coast and large inhabited centres, scarce presence of essential services especially in smaller municipalities, highly developed commuting particularly for students and workers, strong need to develop the tourist service and to transport additional items such as food, mail and medicines.

From the analysis of the results of the first months of experimentation of the DRT service operating in Val Graveglia (Ne), conducted in Chap. 6, policymakers can learn important information (RQ3): the trial was positive overall, attracting numerous users who decided to abandon the use of cars in favour of the new DRT service, such as to justify the permanence of the service even after the pilot.

Furthermore, population of these areas consists mainly of elderly people unfamiliar with the use of technology; thus the app as the only booking option can represent a major problem for the effectiveness of the service. The PTAs should thus insist on the use of the call centre as the main booking tool but, at the same time, stimulate residents to adopt technologically advanced booking methods (app or website) through dedicated training courses.

This study inevitably presents research limitations and future agendas.

As regards the study of results of the DRT service implemented in Val Graveglia (Chap. 6), it was possible to access only partial data referring to the first five pilot months: in the future, the analysis could expand to the other on-call services implemented by AMT in the province of Genoa and, on the basis of data referring to longer periods of service, it would be possible to obtain more robust and reliable indications on the travel behaviour of the population and on strengths and weaknesses of the service.

In conclusion, having established the viability of on-call service as a substitute for FT in low-demand areas, transit providers must find the ideal compromise between two opposing viewpoints: on the one hand, the provision of an economically sustainable service, which, however, only gathers a small number of users due to the medium-high cost of transportation and the strong rigidity of the service. On the other hand, a transportation service that is unprofitable but aims to meet customer demands by offering inexpensive tickets and the most possible flexibility in terms of service ("many-to-many" model).

Currently, a median solution appears to be the cornerstone for an effective deployment of DRT in rural areas, as the Val Graveglia trial attests: the capacity of technical advancement to reduce significantly transportation and technological equipment costs, enabling more flexible and customized services, will be crucial for the future of this tool.

References

Adli, S. N., Chowdhury, S., & Shiftan, Y. (2019). Justice in public transport systems: A comparative study of Auckland, Brisbane, Perth and Vancouver. *Cities, 90*, 88–99.

Ambrosino, G., & Romanazzo, M. (2002). *I servizi flessibili di trasporto per una mobilità sostenibile*. ENEA.

Ambrosino, G., Boero, M., Eloranta, P., Engels, D., Finn, B., & Sassoli, P. (2000). Flexible mobility solutions in Europe through cooperation between operators. In *7th World Congress on ITS CD-proceeding*.

Ambrosino, G., Mageean, J. F., Nelson, J. D., & Romanazzo, M. (2004). Experience and applications of DRT in Europe. Demand Responsive Transport Services: Towards the Flexible Mobility Agency. Rome: ENEA, 139–196.

Attard, M., Camilleri, M. P., & Muscat, A. (2020). The technology behind a shared demand responsive transport system for a university campus. *Research in Transportation Business & Management, 36*, 100463.

Ayland, N. (2000). *Systems for the advanced management of public transport*. SAMPLUS.

Bakker, P., & Van der Maas, C. (1999). *Large scale demand responsive transit systems—A local suburban transport solution for the next millennium?* AVV.

Banister, D. (2018). *Inequality in transport* (p. 272). Alexandrine Press.

BBC. (2023). *Bringing Doctors to the Patients*. From BBC StoryWorks. Retrieved from https://www.bbc.com/storyworks/specials/world-of-possibility/bringing-doctors-to-the-patients/

Behrens, R., McCormick, D., Orero, R., & Ommeh, M. (2017). Improving paratransit service: Lessons from inter-city matatu cooperatives in Kenya. *Transport Policy, 53*, 79–88.

Beiler, M. O., & Mohammed, M. (2016). Exploring transportation equity: Development and application of a transportation justice framework. *Transportation Research Part D: Transport and Environment, 47*, 285–298.

Brake, J., Nelson, J. D., & Wright, S. (2004). Demand responsive transport: Towards the emergence of a new market segment. *Journal of Transport Geography, 12*(4), 323–337.

Burlando, C., & Cusano, M. I. (2014). *Consequences of demographic changes on urban mobility: An overview of ageing in modern societies*. Percorsi di Scienze Economiche e Sociali.

Calabrò, G., Inturri, G., Le Pira, M., Pluchino, A., & Ignaccolo, M. (2020). Bridging the gap between weak-demand areas and public transport using an ant-colony simulation-based optimization. *Transportation Research Procedia, 45*, 234–241.

Campagna, A., & Ambrosi, M. (2013). LIMIT4WEDA results and development. In *11th European Week of Regions and Cities*.

T. Pavanini, *Rural Demand Responsive Transport*, SpringerBriefs in Operations Management, https://doi.org/10.1007/978-3-031-91395-2

Campisi, T., Canale, A., Ticali, D., & Tesoriere, G. (2021). Innovative solutions for sustainable mobility in areas of weak demand. Some factors influencing the implementation of the DRT system in Enna (Italy). *AIP Conference Proceedings, 2343*(1), 090005.

Carrosio, G., Lucatelli, S., & Barca, F. (2018). Le aree interne da luogo di disuguaglianza a opportunità per il paese. In *Le sostenibili carte dell'Italia* (pp. 167–186). Marsilio.

Cole, L. M. (1968). *Tomorrow's transportation: New systems for the urban future* (Vol. 62). US Government Printing Office.

Coutinho, F. M., van Oort, N., Christoforou, Z., Alonso-González, M. J., Cats, O., & Hoogendoorn, S. (2020). Impacts of replacing a fixed public transport line by a demand responsive transport system: Case study of a rural area in Amsterdam. *Research in Transportation Economics, 83*, 100910.

Currie, G., & Fournier, N. (2020). Why most DRT/micro-transits fail – What the survivors tell us about progress. *Research in Transportation Economics, 83*, 100895.

Dicuonzo, B. (2020). *Approccio co-modale e gerarchico nella mobilità attuale e futura per le aree a domanda debole.*

Eloranta, P. (1998). *The Sampo-Project. Results, conclusions and recommendations.* Publications of the Ministry of Transport and Communications.

Enoch, M., Potter, S., Parkhurst, G., & Smith, M. (2004). *Inter-mode: Innovations in demand responsive transport report for Department for Transport and Greater Manchester Passenger Transport Executive.* Department for Transport.

Enoch, M., Potter, S., Parkhurst, G., & Smith, M. (2006, 22–26 January). Why do demand responsive transport systems fail? In *85th Annual Meeting.*

Ferrari, E. (2018). *Servizio di trasporto a prenotazione del bacino di Piacenza: stato di fatto e possibile ampliamento.*

Finn, B. (1996). *Analysis of user requirements for demand responsive transport services.* SAMPO Consortium 1996.

Franco, P., Johnston, R., & McCormick, E. (2020). Demand responsive transport: Generation of activity patterns from mobile phone network data to support the operation of new mobility services. *Transportation Research Part A: Policy and Practice, 131*, 244–266.

Furuhata, M., Daniel, K., Koenig, S., Ordonez, F., Dessouky, M., Brunet, M. E., & Wang, X. (2014). Online cost-sharing mechanism design for demand-responsive transport systems. *IEEE Transactions on Intelligent Transportation Systems, 16*(2), 692–707.

Gallo, G. (2020). *Маршру́тка–Maršrutka.* Retrieved from https://russiaintranslation.com/2020/12/19/marsrutka/

Ghabara, W. (2012). El Mourouj 1 - gouvernorat de Ben Arous - Tunisie Taxi collectif. Own Work. Accessed on https://commons.wikimedia.org/w/index.php?curid=19521375

Goodwill, J., & Carapella, H. (2008). *Creative ways to manage paratransit costs.* USF Center for Urban Transportation Research.

Gössling, S. (2016). Urban transport justice. *Journal of Transport Geography, 54*, 1–9.

Guzman, L. A., Oviedo, D., Arellana, J., & Cantillo-García, V. (2021). Buying a car and the street: Transport justice and urban space distribution. *Transportation Research Part D: Transport and Environment, 95*, 102860.

Häme, L. (2013). *Demand-responsive transport: Models and algorithms.*

Hananel, R., & Berechman, J. (2016). Justice and transportation decision-making: The capabilities approach. *Transport Policy, 49*, 78–85.

Higgins, T. (1976). Demand responsive transportation: An interpretive review. *Transportation, 5*(3), 243–256.

Hirsch, N., & Fredericks, C. (2001). Rural doctors and retention. In *6th National Rural Health Conference.*

Interreg Europe. (n.d.). *Limit4WeDA - Light mobility for weak demand areas.* Retrieved from https://www.interregeurope.eu/good-practices/limit4weda-light-mobility-for-weak-demand-areas

ISTAT. (n.d.). *Codici statistici delle unità amministrative territoriali: comuni, città metropolitane, province e regioni.* Retrieved from https://www.istat.it/it/archivio/6789#:~:text=Dal%20 20%20febbraio%202021%2C%20con,%C3%A8%20pari%20a%207.904%20unit%C3%A0

Jain, S., Ronald, N., Thompson, R., & Winter, S. (2017). Predicting susceptibility to use demand responsive transport using demographic and trip characteristics of the population. *Travel Behaviour and Society, 6*, 44–56.

Javisbg618 (2021). Ruta 15 como principal transporte público que atraviesa la colonia Olivar del Conde. Own Work. Accessed on https://commons.wikimedia.org/w/index. php?curid=110696415

Jennings, G. (2015). Public transport interventions and transport justice in South Africa: A literature and policy review. In *Southern African Transport Conference.*

Jokinen, J. P., Sihvola, T., & Mladenovic, M. N. (2019). Policy lessons from the flexible transport service pilot Kutsuplus in the Helsinki Capital Region. *Transport Policy, 76*, 123–133.

Kanyama, A. (2004). Public transport in Dar es Salaam, Tanzania: Institutional challenges and opportunities for a sustainable transportation system. Totalförsvarets forskningsinstitut, Institutionen för miljöstrategiska studier.

Karim, M. M.. *Own work, GFDL.* Retrieved from https://commons.wikimedia.org/w/index. php?curid=4313089

Karner, A., London, J., Rowangould, D., & Manaugh, K. (2020). From transportation equity to transportation justice: Within, through, and beyond the state. *Journal of Planning Literature, 35*(4), 440–459.

Karner, A., Bills, T., & Golub, A. (2023). Emerging perspectives on transportation justice. *Transportation Research Part D: Transport and Environment, 116*, 103618.

KFH. (2008). *Guidebook for measuring, assessing, and improving performance of demand-response.*

King County (2025). Vanpool. Accessed on https://kingcounty.gov/en/dept/metro/travel-options/ vanpool-and-vanshare/vanpool

König, A., & Grippenkoven, J. (2020). The actual demand behind demand-responsive transport: Assessing behavioral intention to use DRT systems in two rural areas in Germany. *Case Studies on Transport Policy, 8*(3), 954–962.

Lave, R. E., & Mathias, R. G. (2009). Paratransit systems. *Transportation Engineering and Planning, 1*, 207–245.

Laws, R., Enoch, M., Ison, S., & Potter, S. (2009). Demand responsive transport: a review of schemes in England and Wales. *Wales. Journal of Public Transportation, 12*(1), 19–37.

Litman, T. (2002). Evaluating transportation equity. *World Transport Policy & Practice, 8*(2), 50–65.

Lunardon, D. (2011). *C101019-ProgettazioneAutobusChiamata_Vers_4.* Retrieved from http:// www.agenda21laghi.it/download/Quaderno%205%2020Proposta%20di%20autobus%20 a%20chiamata.pdf

Maffii, S., Sitran, A., Brambilla, M., Angelo, M., Mandel, B., & Schnell, O. (2012). *Bigliettazione integrata per i servizi di trasporto passeggeri a lunga distanza.*

Mageean, J., & Nelson, J. D. (2003). The evaluation of demand responsive transport services in Europe. *Journal of Transport Geography, 11*(4), 255–270.

Marchioro, C. (2018). *Dinamiche socio-economiche nelle aree interne della Liguria.* ASITA.

Martens, K. (2012). Justice in transport as justice in accessibility: Applying Walzer's 'spheres of justice' to the transport sector. *Transportation, 39*, 1035–1053.

Martens, K. (2016). *Transport justice: Designing fair transportation systems.* Routledge.

Mastronardi, L., & Romagnoli, L. (2020). Community-based cooperatives: A new business model for the development of Italian inner areas. *Sustainability, 12*(5), 2082.

Matsuhita, S., Yumita, S., & Nagaosa, T. (2022, October). A proposal and performance evaluation of utilization methods for tourism of a demand-responsive transport system at a rural town. In *In 2022 IEEE 25th international conference on Intelligent Transportation Systems (ITSC)* (pp. 2920–2925). IEEE.

Mattes, (2007). Medium sized Songthaew built by HINO Motors - seen in Sakon Nakhon, Thailand. Accessed on https://commons.wikimedia.org/w/index.php?curid=1728961

Mfinanga, D., & Madinda, E. (2016). Public transport and daladala service improvement prospects in Dar es Salaam. In *Paratransit in African cities: Operations, regulation and reform* (pp. 155–173). Routledge.

Mishra, S., Brakewood, C., Golias, M. M., Aravind, A., Venthuruthiyil, S. P., & Sharma, I. (2022). *Connecting demand response transit with fixed service transit (No. RES-2021-03)*. Department of Transportation.

NADTC. (2014). *ADA guide for rural demand-responsive transportation service providers*. Easter Seals Project Action. Retrieved from https://www.nadtc.org/wp-content/uploads/NADTC-ADA-Guide-for-Rural-Demand-Response-Providers.pdf

Nelson, J. D., Ambrosino, G., & Romanazzo, M. (2004). *Demand responsive transport services: Towards the flexible mobility agency.*

Nelson, J. D., Wright, S., Masson, B., Ambrosino, G., & Naniopoulos, A. (2010). Recent developments in flexible transport services. *Research in Transportation Economics, 29*(1), 243–248.

Nyga, A., Minnich, A., & Schlüter, J. (2020). The effects of susceptibility, eco-friendliness and dependence on the consumers' willingness to pay for a door-to-door DRT system. *Transportation Research Part A: Policy and Practice, 132*, 540–558.

Panikkar, B., Ren, Q., & Bechthold, F. (2023). Transportation justice in Vermont communities of high environmental risk. *Sustainability, 15*(3), 2365.

Papanikolaou, A., & Basbas, S. (2021). Analytical models for comparing demand responsive transport with bus services in low demand interurban areas. *Transportation Letters, 13*(4), 255–262.

Papanikolaou, A., Basbas, S., Mintsis, G., & Taxiltaris, C. (2017). A methodological framework for assessing the success of demand responsive transport (DRT) services. *Transportation Research Procedia, 24*, 393–400.

Pavanini, T. (2023). Demand responsive transport in Italian rural areas: State of the art of technical characteristics and level of innovation of 35 case studies. *Revista de Estudios Andaluces, 45*, 123–145. https://doi.org/10.12795/rea.2023.i45.07

PFA. (2017). *Accordo di programma quadro Regione Liguria - "AREA INTERNA – Valli dell'Antola e del Tigullio".*

Phun, V. K., & Yai, T. (2016). State of the art of paratransit literatures in Asian developing countries. *Asian Transport Studies, 4*(1), 57–77.

Rawls, J. (1971). An egalitarian theory of justice *Philosophical Ethics: An Introduction to Moral Philosophy*, 365–370.

Regione Lazio. (2013). *Light mobility and information technologies for weak demand areas.*

Regione Liguria. (2020). *Relazione di Avanzamento Annuale. Accordo di Programma Quadro "Area Valli dell'Antola e del Tigullio", 13 giugno 2020.*

Regione Liguria. (n.d.). *PTCP - Assetto Insediativo e Aree Carsiche*. Retrieved from http://srv-carto.regione.liguria.it/geoviewer2/pages/apps/geoportale-tecnico/index.html?id=1461

Rissanen, K. (2016). *Kutsuplus - Final report*. Helsinki Regional Transport Authority: Helsinki Regional Transport Authority (HSL) Publications.

Roschlau, M. W. (1981). *Urban transport in developing countries: The Peseros of Mexico City.*

Sagaris, L., & Tiznado-Aitken, I. (2018). *Walking and gender equity: Insights from Santiago Chile (No. 18-05195).*

Scholliers, J. (2002). *INVETE, Final report.*

Scholliers, J., & Valenti, G. (2002). Evaluation of an open intelligent in-vehicle terminal for regular and flexible public transport services. In *9th World Congress on Intelligent Transport Systems ITS America*. ITS Japan, ERTICO (Intelligent Transport Systems and Services-Europe).

Schumacher, O. (2020). *DB Medibus—The Mobile Clinic*. Deutsche Bahn AG.

Sheller, M. (2018). *Mobility justice: The politics of movement in an age of extremes.* Verso Books.

SNAI. (2013). *Strategia nazionale per le Aree interne: definizione, obiettivi, strumenti e governance. Accordo di Partenariato 2014-2020.*

Takeuchi, R., Okura, I., Nakamura, F., & Hiraishi, H. (2003). Feasibility study on demand responsive transport systems (DRTS). In *5th Eastern Asia Society for Transportation Studies Conference.*

Talley, W. K., & Anderson, E. E. (1986). An urban transit firm providing transit, paratransit and contracted-out services: A cost analysis. *Journal of Transport Economics and Policy,* 353–368.

UVAL. (2014). *A strategy for inner areas in Italy: Definition, objectives, tools and governance.* Casavola, Paola.

van Holstein, E., Wiesel, I., & Legacy, C. (2022). Mobility justice and accessible public transport networks for people with intellectual disability. *Applied Mobilities, 7*(2), 146–162.

Vanoutrive, T., & Cooper, E. (2019). How just is transportation justice theory? The issues of paternalism and production. *Transportation Research Part A: Policy and Practice, 122,* 112–119.

Verlinghieri, E., & Schwanen, T. (2020). Transport and mobility justice: Evolving discussions. *Journal of Transport Geography, 87,* 102798.

Vij, A., Ryan, S., Sampson, S., & Harris, S. (2020). Consumer preferences for on-demand transport in Australia. *Transportation Research Part A: Policy and Practice, 132,* 823–839.

Walzer, M. (2019). *Spheres of justice.*

Walzer, M. (1983). Spheres of Justice: A Defense of Pluralism and Equality. New\brk: Basic Books.

Wang, C., Quddus, M., Enoch, M., Ryley, T., & Davison, L. (2015). Exploring the propensity to travel by demand responsive transport in the rural area of Lincolnshire in England. *Case Studies on Transport Policy, 3*(2), 129–136.

Weckström, C., Mladenović, M. N., Ullah, W., Nelson, J. D., Givoni, M., & Bussman, S. (2018). User perspectives on emerging mobility services: Ex post analysis of Kutsuplus pilot. *Research in Transportation Business & Management, 27,* 84–97.

Weicker, T. (2020). Marshrutka (in)formality in southern Russian cities and its role in contentious transport policies. *Geoforum, 136,* 262–272.

Wilson, N. H., Sussman, J. M., Goodman, L. A., & Higonnet, B. T. (1969). *Simulation of a computer aided routing system (CARS).* Institute of Electrical and Electronics Engineers (IEEE).

Woolf, S. E., & Joubert, J. W. (2013). A people-centred view on paratransit in South Africa. *Cities, 35,* 284–293.

Yen, B. T., Mulley, C., & Yeh, C. J. (2023). Performance evaluation for demand responsive transport services: A two-stage bootstrap - DEA and ordinary least square approach. *Research in Transportation Business & Management, 46,* 100869.

The manufacturer's authorised representative in the EU is Springer
Nature Customer Service Centre GmbH, Europaplatz 3, 69115 Heidelberg,
Germany. If you have any concerns regarding our products, please
contact ProductSafety@springernature.com

Printed and bound by CPI Group (UK) Ltd, Croydon, CR0 4YY

29/04/2026

02099459-0014